ons would be to tetr
ted along the [111]
ally occupied.  As
60°) distortion for
< 1) CoO has atomi

much of the code w
the intended messag
quired degree of rel
feedback applies to
transmission system i
is a function of bot
formation available
results of previous t

agona
s,
shown

$_0/N_o$.  If we assum

FeO,
c mom

-53-i  Bounds on the error
asymptotic relation
[1 of error and the ave
achieve it are deriv
transmission.  It is s
a sequential-sphere
(decision-feedback),
packed block code,
capacity, than the
a fixed-constraint-l
tured that it is at le
izable decision-fee
however, that asym
is possible with an i
transmission system

$r(t_1 - t_2) = \dfrac{1}{2N_o}$ { 1

968

rm the logarithm o
ived data are, after

$"[\psi_2] = \text{Re } \beta_2^* \left[ \int_0^T \right.$

se of higher-order feedback, even larc
he symmetry is rhc tained with continuc
prepared.  Recent  The latter point is il
information-feedbac
m through the latti which assumes the e
channel.  Its error e
ugh these inclusion transmission rate, v
hey probably do no dichotomy" exponen
length dichotomy" a
ince $\Delta_{LS} > \Delta_{JT}$ in other systems mentic
rtions in tetragona is zero.)  Bounds are
78   and $Cu[Cr_2]O_4$  straint length at cap
of the system reveal
O and CoO, noncol length required to a
10^{-15}

aniso

$-\dfrac{R_{10}\sigma_\delta}{1 + R_{10}}$   mbohe
neutro
ce, bu
s may

of such a detector
r of Eq. (40) (for al

t alter
FeO

Ni[Cr
c/a =
linear

W9-CBP-415

# INSIDE THE MACHINE

Also by

**MEGAN PRELINGER**

# ANOTHER SCIENCE FICTION:

Advertising the Space Race 1957–1962

Art and Invention in the Electronic Age

# INSIDE THE MACHINE

## MEGAN PRELINGER

W. W. NORTON & COMPANY

NEW YORK · LONDON

All illustrations courtesy of the Prelinger Library, with the exception of *figs.* *1.10* (courtesy of the Hagley Museum and Library); *3.7* (reproduced with permission, © 1949 Scientific American, Inc., all rights reserved); *3.9* (courtesy of Elizabeth Staley and the Schenectady Museum of Innovation and Science); *5.9*, *6.9*, *6.18*, and *8.19* (courtesy of Willi Baum); *5.14* (courtesy of David Rumsey/Pulsa); and *9.3* (courtesy of Eric Fischer).

For information about permission to reproduce selections from this book, write to Permissions, W. W. Norton & Company, Inc., 500 Fifth Avenue, New York, NY 10110

For information about special discounts for bulk purchases, please contact W. W. Norton Special Sales at specialsales@wwnorton.com or 800-233-4830

Manufacturing by RR Donnelley, Shenzhen
Book design by Chin-Yee Lai
Production manager: Anna Oler

Library of Congress Cataloging-in-Publication Data

ISBN 978-0-393-08359-0

W. W. Norton & Company, Inc.
500 Fifth Avenue, New York, N.Y. 10110
www.wwnorton.com

W. W. Norton & Company Ltd.
Castle House, 75/76 Wells Street, London W1T 3QT

1 2 3 4 5 6 7 8 9 0

Dedicated to Z.

## CONTENTS

INTRODUCTION ... 10

ONE • THE ATOM, THE PLANET, AND THE TUBE ... 29

TWO • TUBES THAT SEE: CATHODE-RAY TUBES ... 45

THREE • COLD ROCK, WARM LIFE: CRYSTALS ... 67

FOUR • TRANSISTORS AND CIRCUIT SYMBOLS ... 85

FIVE • CIRCUIT BOARDS AND THE MATRIX ... 103

SIX • AUTOMATIC AND DIGITAL: THE EMERGENCE OF COMPUTING ... 123

SEVEN • VISIBLE LANGUAGE ... 145

EIGHT • THE FURTHEST HORIZON: SPACE ELECTRONICS ... 171

NINE • BIONICS, A PROLOGUE TO TRANSHUMANISM ... 195

NOTES ... 222

REPOSITORIES CONSULTED ... 243

REFERENCES FOR FIGURES ... 244

ACKNOWLEDGMENTS ... 252

INDEX ... 255

# INSIDE THE MACHINE

# INTRODUCTION

The New Deal has restricted the hours of labor and thereby has introduced a tremendous amount of New Leisure. . . . Sooner or later this "Niagara of leisure" will have to be absorbed. . . . A large part will be devoted to such electronic entertainment as radio, sound-pictures, home talkies, phonograph reproduction, synthetic musical instruments, and eventually, television.

—*Electronics* magazine editorial, September 1933

Engineers of the early twentieth century held an invisible new power in their hands: they could control the flow of electrons. Harnessing electrons to serve human purposes was a technological leap that would change the shape of the century. Radio would be the first worldwide application of this new power, followed by television and later telecommunications and computer networks, with a thousand smaller steps in between. Today's micro devices, tucked imperceptibly into our bodies and our personal tools, are the direct descendants of those discoveries.

Today's cloud computing facilities actively defy their physicality, while the smallest devices have shrunk to the scale of the electron. Computing culture embraces not-being-seen. In contrast, for most of the twentieth century electronics were new devices that benefited from visibility. The component parts of electronic devices—tubes, transistors, and circuits—went through a bell-shaped cycle of emergence: they became steadily more widely seen from the 1910s through the 1930s, then popular and familiar in the 1940s and 1950s. One by one, components entered the public stage and took their places as new artifacts in the cultural firmament.

The devices that they made possible, such as radio, television, and computers, led the visibility of electronics. But the emergence of components had more in common with the systems that those components enabled—telecommunications, transportation, and information. Components and systems both developed outside the view of most people who would use them. In order for them to be introduced and understood, they were placed at the

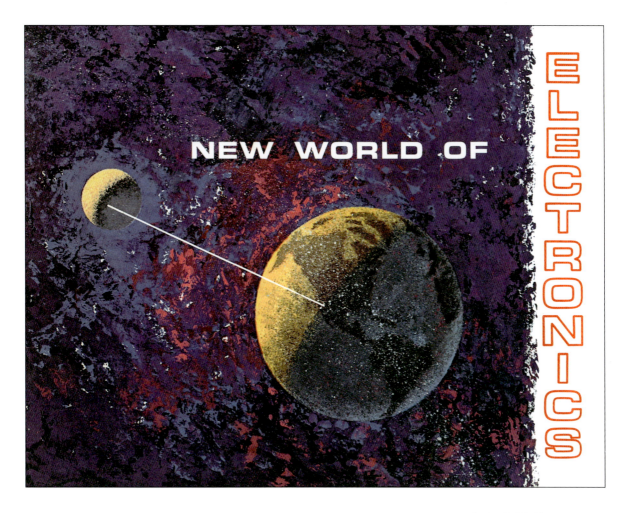

NEW WORLD OF ELECTRONICS

center of advertising and literature campaigns that drew on the talents of graphic artists to convey otherwise invisible developments. The large mainframe computers of the 1960s were the turnaround point of these campaigns for visibility. From the 1960s onward, electronics shrank in size and retreated from view, heading toward today's invisibility.

That this period of visibility happened is remarkable. Components themselves were not marketed to the general public, neither were computers in their early decades. The flow of electrons could not be seen, nor could anything as abstract as the process of computation or the emergence of information systems. Each component was a successive invention that emerged from a laboratory and from there became part of a closed system. They reached our pockets and living rooms when they became parts of a new whole: the new radio! the new TV! Yet they did come to light, made vivid on the pages of magazines, catalogs, and books aimed at a techno-savvy readership. The medium of their reveal was art made for industry; the result was a dramatic intersection of art and technology.

This book is a history of electronics that explains technology through the lens of art. It also asks the question: What cultural history of electronics can be extrapolated from a close look at the associated commercial graphic art?

The result is a guided tour through the commercial and advertising art that was created when the labors of engineers met the inspiration of artists. At its heart is a technological story of the development of electronic components from laboratory to tabletop. The medium of exploration is the artwork through which new components were introduced, contextualized, and promoted. What's revealed is art's ability to touch the intangible and render it visible.

Fig. i.2:
ELECTRICITY AND THE
SOLAR SYSTEM: THE ARTIST
MAXFIELD PARRISH CREATED
AN AD CAMPAIGN FOR
GENERAL ELECTRIC IN 1917
TO PROMOTE THE COMPANY'S
EDISON MAZDA LAMP BRAND.

## COMMERCIAL ART FOR THE ELECTRONICS INDUSTRY: THE FIRST FIFTY YEARS

The printed literature written and illustrated in the mid-twentieth century to explain electronics provides the raw material for this book. These documents, made primarily for industry and secondarily for the general public, are part of a long tradition of commercial art that stretches back to nineteenth-century lithography. The discipline of graphic art—art created for reproduction—has its origins in the printing boom of the nineteenth century. Early twentieth-century commercial art was a fertile ground for the lush illustration of the brand-new hardware of everyday life, such as the lightbulb.

The mid-century era of this book followed directly from the preceding fifty years of invention and illustration. The humble lightbulb stepped into that tradition when it hit the market in the 1880s. It was the first mass-marketed electrical component. As a stand-alone invention it became the subject of artistic attention, sponsored by manufacturers, to introduce, promote, and contextualize it for the public. The synchronous development of the telephone was a similarly tangible outcome of the new-found ability to control electricity within circuits. The respective laboratories of Alexander Graham Bell (later, AT&T's Bell Telephone Laboratories) and Thomas Edison (whose Edison General Electric company became General Electric) were neighboring epicenters of discovery in methods of harnessing and applying the power of electricity. GE commissioned the noted illustrator Maxfield Parrish to create a series of promotions for the company's Edison Mazda lamp (a lightbulb); *fig. i.2* is one of many results. The slogan "His Only Rival" (referencing the sun) promotes the bulb in distinctly planetary terms. Both this image and *fig. i.3* foreshadow decades of connections that would be drawn between electronic technologies and the solar system.

These inventors had historical contemporaries whose work was of equal technological significance even if it was less well known. Nikola Tesla, for example, developed alternating current (the AC of AC/DC), an electrical technology that is as crucial to today's everyday infrastructure as any of Edison's inventions. However, Tesla's discoveries did not lead to either a public company or a range of industrial and consumer products bearing his name. As a result, the memory of his work was dispersed for a century. The circumstance of Tesla's legacy points out that the total universe of electronic experimentation and invention is much broader than what is represented here. As an introduction to electronic technology, this study is confined to the particular trail of evidence found at the intersection of commerce and graphic art.

### ELECTRICITY CHANGES THE SHAPE OF SPACE-TIME

The best-known inventions by Edison and Bell, the lightbulb and the telephone respectively, each transformed the human sense of scale. The telephone's effect on scale was to carry voices hundreds of miles. The lightbulb's effect was to change the shape of the day. It allowed people to see further at night, but even more profound was the ripple effect of that expanded vision, which was to change how people experienced time. Electric light expanded the usable number of hours in the day, making the day functionally longer. Both technologies were depicted in commercial art in the illustrative style that characterized magazine and pamphlet literature at the turn of the twentieth century. *Fig. i.3* makes a mathematical point about the shift in scale that the telephone made possible, but it also casts that point in rounded visual terms: the gentle ovals of planetary motion offer an understated frame of reference for the telephone network.[1]

Into that environment appeared the vacuum tube, the first electronic device, at the dawn of the twentieth century. Its invention represented a quantum technological leap beyond its immediate predecessor, the lightbulb. Its

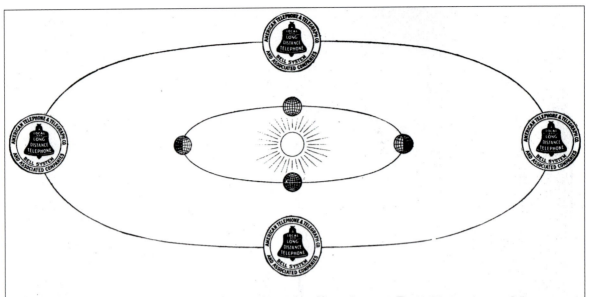

*Comparison of the Distance Traveled by Earth and Bell Telephone Messages*

# The Orbit of Universal Service

shape was similar to that of a lightbulb, but inside its workings a new dimension had opened up: the invisible flow of electrons could now be controlled, and their *electronic* power would unleash a torrent of technological change. Companies that wanted to explain what a radio was, or how television worked, or sell their particular tube to a manufacturer, had to hire artists who could extrapolate a visual identity from a hidden, if essential, new tool.

**MID-TWENTIETH-CENTURY COMMERCIAL ART**

In the 1930s, vacuum tubes amplified the radio decade and turned it up loud. As radio technology was fairly new to general home and business use in 1930, the industry experienced growth even during the Great Depression. It barely looked like growth, but any development in the field represented an expansion beyond the near nonexistence of radios for home and business a decade earlier.[2] The 1930s also saw the early integration of electronics with automated systems, such as those that governed telephone switches and streetcar railway networks, turning those mechanical and electromechanical systems into electronic systems. *Electronics* magazine, founded in 1930 to report on the development of these new technologies, managed to maintain a general spirit of optimism throughout its early years in spite of the global economic depression. The developments in radio, telecommunications, and transportation during those years prepared the electronics industry to run with the subsequent demands placed upon them by wartime. By the end of the decade, American industry was on the cusp of an enormous war-related growth spurt.

As electronics expanded, the changing political environment in Europe caused a ripple effect that would help shift commercial art away from its origins in the illustrative tradition. Europe was on edge in the 1930s. Many European artists, including highly trained practitioners of modern art, began to flee persecution and looming war to make new lives for themselves in the United States. These immigrant artists infused the American artistic community with new energy and new ideas, bringing with them techniques from European schools of artistic thought. The Italian Futurist movement had liberated printed words from their conventions in the 1909 *Futurist Manifesto*; French surrealists had begun in the 1920s to paint the dream-like "reality" behind, or above, everyday life. Russian Constructivists found strength in the irreducible and blocklike essentials of line and form, and the Bauhaus school, originally German, embraced geometric shape and collage, among other modernist strategies. All of these traditions, and others, found their way into the repertoire that commercial artists drew upon to depict invisible and intangible developments in electronic technology.

Perhaps even more scientists than artists fled Europe in the prewar years; they stocked American laboratories and universities with many of the

Fig. i.4:
RCA (*SCIENTIFIC AMERICAN*, 1960).

**BEYOND THE BOUNDARIES OF KNOWLEDGE...**

greatest theoretical and experimental minds of the century. Pure laboratory research took longer to produce visible fruit than did graphic art, but both would contribute enormously to their respective domains of mid-century art and science in the U.S.

The emerging dominance of the new electronics industry also had its effect on the dynamic between commercial art and technology. Electronic technology relies on an invisible process, the controlled movement of electrons. Electricity was essentially invisible too, but it could be felt, and it could be represented as sparks and bolts. The utter invisibility of the electron put it in a new dimension of (in)tangibility, demanding different strategies of representation from artists. Lastly, electronic technology enabled the swift development of a sprawling new suite of industries and industrial processes that transformed American business and industry in a few short years. When the computer industry emerged late in this sequence of events, it expressed the combined strengths of a number of fast-developing antecedent technologies. This proliferation of new businesses had a predominant interest in selling to one another (rather than to the general public), as many of them manufactured only component parts, not finished products. Business-to-business literature expanded quickly to keep up.

The artwork at the center of this book emerged from this period of colliding forces. The American graphic tradition of commercial art met European modernism at the same time that it met a new, vibrant, and seriously abstracted subject. When World War II came, commercial art took on a new significance as its subject facilitated an immense industrial boom. Business-to-business communication, both advertising and pamphlet literature, became a forum for some of the most sophisticated modernist graphic art of the twentieth century.[3] Within the electronics industry—and every place it touched—this forum explained hard-to-understand technologies, conveyed the systemic impact of these technologies, and convened an emerging whole out of a multipart industry.

**"ELECTRONICS: TECHNIQUES FOR A NEW WORLD 1"** In July 1943, the editors of *Fortune* magazine declared the year of electronics. It would be the year that electronics developed for wartime applications would become increasingly integrated with domestic markets, and *Fortune* observed the anticipated impact with a splashy theme issue.

*Fortune* declared electronics "a lever on industry" and predicted that the $4 billion war business would generate "a postwar industrial revolution."[4] They were right, although technology moved so fast that their foresight

*Fig. i.5:*
RCA (*SCIENTIFIC AMERICAN,*
1960).

**THE VIEW THROUGH THE 10-MICRON WINDOW**

began to fall short by 1948, as we will see. To illustrate the cover article, the magazine's art directors commissioned a glossy four-panel insert titled "Electronics: Techniques for a New World 1," a vivid panorama of the major electronic components of the day (*fig. i.6*).[5] In its range and diversity, this poster expresses several characteristics of 1940s' art that was created to explain and promote electronics. It was not the first remarkable artwork created for its purpose, but it was by far the largest and most elaborate.

Artwork created to accompany *Fortune* articles complemented neighboring advertising artwork and often surpassed it in scope and quality. Yet it wasn't wholly a breed apart: *Fortune* was sponsored, for-profit education, directed at the decision-makers within business and industry, written and illustrated to educate them about trends. The *Fortune* magazine of the mid-twentieth century was a glittering diamond atop a mountain of less well-funded labors. The editors and art directors at the magazine set world standards for

*Fig. i.6:*
"ELECTRONICS: TECHNIQUES
FOR A NEW WORLD 1:
A *FORTUNE* COLLAGE"
(*FORTUNE*, NOVEMBER 1943).

the production of sophisticated artwork in support of deeply researched articles. There is no contemporary analogy; *Fortune*'s quality stands alone in the past century's history of integration of image with message. Published in a large format (10½" x 13") and printed in full color on thick, matte-finish paper, an issue of *Fortune* from 1943 was a four-pound package of creamy heft in service of American commerce. And *Fortune* was excited about electronics.

The poster depicts the electronic component of the day, vacuum tubes. Eight types of tubes are shown, and the names they are given reveal the intimate relationship between big industry and the process of invention and marketing of new components. During the 1930s and 1940s, the electronics industry was ruled by three titans: General Electric, Bell Laboratories, and RCA (the Radio Company of America). These companies were followed closely in their scope of influence by IBM, originally a manufacturer of mechanical and electromechanical business machines, whose dominance would expand in the postwar years. The *Fortune* article was researched in close consultation with General Electric—all the tubes in the poster are GE tubes. Other manufacturers, especially RCA and Bell Laboratories, produced their own similar versions.

The poster's lead designer, Peter Vardo,[6] was a prominent commercial artist who was noticed in the world of fine art. His work for the magazine during the 1940s would later be cited as having influenced Jackson Pollock, and would be analyzed by art historians in the decades to come.[7] He collaborated on the design with Warren Stokes and Antonio Petruccelli, another highly regarded commercial artist. Petruccelli's finely detailed paintings would appear on the covers of twenty-five issues of *Fortune* during his tenure with the magazine.[8]

The three designers compiled illustrative and photographic elements from a range of sources. From the professional photographer Frederic Lewis came the collage's background cloud photograph. The freelance commercial photographer Fritz Goro was commissioned to make photographic portraits of the eight featured vacuum tubes. From the front-line work of *Life* magazine photographer Frank Scherschel, who was in Europe at the time documenting the war, came the American bombers in flight. And then industry chimed in: the x-ray of a human hand was provided by General Electric and the welding process photograph (top right) by Westinghouse. The photograph of the klystron tube was provided by the *Journal of Applied Physics*. The other, illustrative elements were undoubtedly created by Vardo and Petruccelli. The illustrations add color and depth to the collage, and contribute to the sense of the scientific context necessary for understanding electronics.

This poster exemplifies the source material that inspired this book. In it, visual elements from industry, science, war, and from professional artists combine to create a new work, one that refers outward to the realm of fine art as well as to the poster's central mission of educating a techno-savvy segment of the public and leaders of business and industry. Organic elements—clouds, humans, even a butterfly—soften the image and subtly position the new technology as a shifter between organic and mechanical systems. The collage itself expresses a modernist approach to graphic design: the style of alignment and overlap of images draws on the style cultivated by artists in the Bauhaus school, as well as established avant-garde artists. The poster's format references a scientific diagram, with its linear progression from left to right tracked by a banded spectrum of color. The very strong mediating force of commerce shapes the relationship between art and technology here, as everywhere in this book. In the book's nine chapters, illustrations are points of departure for a basic technological explanation of how electronics components developed in the twentieth century and how they formed devices, systems, and finally networks. They also help the reader learn to look for less mediated, or differently mediated, intersections of art and technology in the world we live in.

**THE DIALOGUE BETWEEN ART AND SCIENCE, AND ART AND TECHNOLOGY** The *Fortune* poster was not discussed by art historians in its own time. Our contemporary dialogues about art and science, as well as those about art and technology, continue a tradition that only commenced in the late 1940s. Similarities between the geometric forms of modern art and those of cell and crystal structures were first investigated by György Kepes, a Hungarian-born artist and theorist of modern art who emigrated to the U.S. in 1937. He founded the Center for Advanced Visual Studies at MIT in 1947 to study these interrelationships, forming a channel to connect microphotography—photography through microscopes—to the practice of contemporary art. In 1956 he produced an exhibit and accompanying book, *The New Landscape in Art and Science,* that first paired modern painting with scientific photography. The impact of Kepes's work on the art world was strongest in the 1960s, contemporaneous with the birth of the long-running group of artist and scientist collaborators based at Bell Laboratories, Experiments in Art and Technology (EAT). At Yale University a few years later, the group Pulsa was formed of engineer-artists who published the technical reports of their work in engineering magazines even as their electronic artwork gained recognition in the world of fine art.

Contemporary discussions about the relationship between art and science tend to draw upon this turn in the 1960s toward direct collaboration between artists and scientists, and artists and inventors, and to focus on technology-based art forms, such as video art and electronic music. Historically, the 1960s was also a decade when academic dialogues about art and science brought close attention to the impact of microphotography on architecture, design, and fine art. In the preceding decades there had been little developed discourse about the relationship between either art and science, or art and technology, beyond Kepes. Then, in 1969, the journal *Leonardo* was founded by the engineer and artist Frank Malina specifically to address these domains.

This book steps back in time from these well-studied beginnings to look at the relationship between art and both science and technology, based primarily on the work of commercial graphic designers. I use the phrase "art and technology" when referring generally to the works explored in this book, as they primarily depict inventions that scientific research enabled rather than the pure science. In many cases, artists making new works about technology reached to the science behind those technologies as a visual and conceptual resource, especially in the case of products created by organizations like Bell Laboratories and General Electric, which were research science institutions as well as commercial manufacturers. In reference to those specific cases—and there are many of them—the familiar phrase "art and science" is appropriate. However, because "art and science" occurs here only in attempts to explain or promote technology, and not the other way around, "art and technology" is more appropriate as an umbrella term for the works explored here.

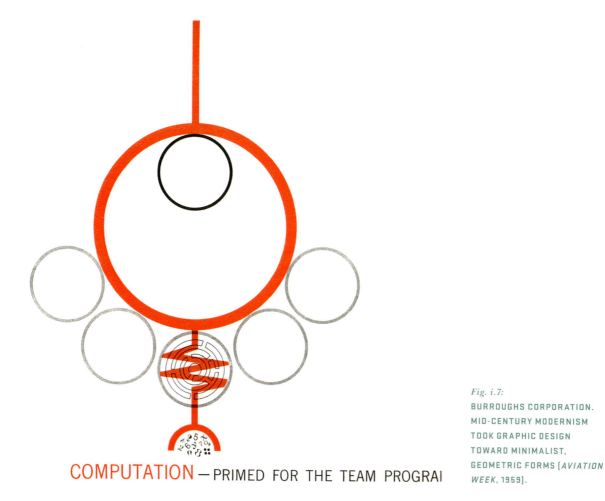

COMPUTATION — PRIMED FOR THE TEAM PROGRAI

*Fig. i.7:*
BURROUGHS CORPORATION. MID-CENTURY MODERNISM TOOK GRAPHIC DESIGN TOWARD MINIMALIST, GEOMETRIC FORMS (*AVIATION WEEK*, 1959).

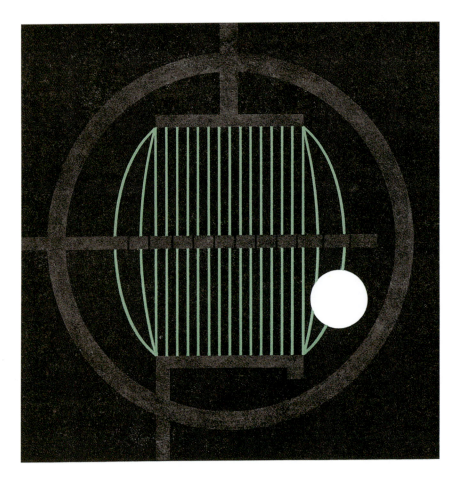

Prior to the 1960s, the business environment was the single most important facilitator of art-about-technology. Businesses *needed* art to describe and explain new technologies to their audience. In the case of hundreds of component manufacturers, they needed to promote their wares to potential purchasers first and only secondarily to the public. As *Industrial Marketing* observed about the electronics industry in 1964, it was "self-oriented." The meta-magazine of corporate promotions also noted that the electronics industry invested more in magazine advertising than other formats because the technology moved too fast for traditional business-to-business formats such as quarterly or annual catalogs.[9]

Yet even trade periodicals were widely read by the interested public, many of whom found magazines like *Radio-Electronics* useful in orienting themselves to possible future careers or hobbies in the field. The general interest magazine *Scientific American* was the principal sponsor, outside the business community, of commissioned artworks that explained and illustrated scientific and technical concepts for the public.

The degree of mediation afforded by commerce is more variable than might be expected. Most artists worked freelance with ad agencies that matched them to clients. Some were sent to laboratories to research their subjects firsthand; most worked with materials sent to them by agencies or other intermediaries. A few artists worked in-house at electronics firms, hand in hand with scientists. This set of interrelationships between art and technology, though conducted within a business environment, was often rich and multi-chambered.

**MODERNISM, AND MODERNITY**

Artists bridged the gap between invention and understanding, between business and industry, and between technology and the public. Electronics embodied twentieth-century technological *modernity* in a way that harmonized with twentieth-century traditions of graphic *modernism*. Graphic modernism developed partly in response to an increasing sense of alienation in everyday life, an alienation that was rooted in the impact of urbanism and technology. Many modernist strategies of geometric pattern-making and integration of typographic elements, among others, found unlooked-for twins and mirrors in the world of electronics. The resulting artworks contributed to a conversation within the realm of the underlying science itself: whether to hew to a "natural science" (organic) model of physical phenomena (exemplified by the invocation of the solar system displayed in *figs. i.1*, *i.2*, and *i.3*) or a mechanistic, technology-oriented model, suggested in the fragmented works of *figs. i.4* and *i.5*.

The artists whose works are featured in this book drew on the full range of illustrative and graphic styles, in addition to modernism. While many were immigrants coming from international traditions, others were trained in domestic schools of traditional illustrative graphic art. In many situations, between the invention of the vacuum tube and the space race of the 1960s, conventional

*Fig. i.9:*
STYLIZED DEPICTION OF A SCULPTURE THAT WAS INSTALLED IN RCA'S NEW YORK EXHIBITION HALL FROM 1947 THROUGH THE 1970S. THE SCULPTURE FEATURED ELECTRONIC COMPONENTS ATTACHED TO A SPIRAL RAMP THAT FLOWED FROM PLANET EARTH TOWARD THE EXHIBIT FLOOR (ALLIED RADIO CATALOG, 1951).

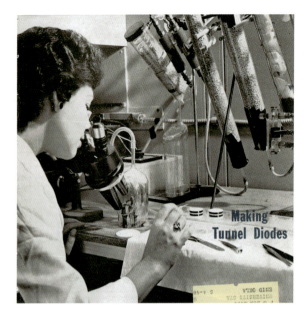

*Fig. i.10:*
A WORKER MAKES DIODES
(*ELECTRONICS*, 1959).

illustrative representation was the most common technique for communicating about electronics. The last two chapters of this book, those on space electronics and biomedical electronics, include many artworks that draw on science-fiction visual vernaculars among other strategies. At the same time, graphic modernism kept step with the more abstract realms of signification offered by the new technologies. Together this range of styles introduced electronics to the public eye.

By the 1970s, with the ascendance of photography, the context for commercial art had once again changed. More significantly, the 1960s marked the end of the era of the electronic component per se. From the 1930s through the 1960s, components themselves were new actors: objects in the material culture of the industrialized world that were valued for their transformative use but were also understood and appreciated as "things" in and of themselves. Commercial artwork interpreted these new things as they entered our compendium of understood artifacts. By the 1960s, room-sized mainframe computers had swallowed these "things" within their big bodies. Their blunt physicality became the new visual reference point for understanding "electronics." Then in the 1970s, the development of microcircuits took electronics sharply in the other direction, returning them to the scale of particulate matter, smaller than the grains of sand from which they had come.

This book treats a corpus of mid-century artworks as double doors: each figure offers a perspective on the history of graphic art and an entry point to a history of technology. The interweaving of these two perspectives yields its cultural historical strategy. The journey from the vacuum tube to transistor to circuit board was a series of successive steps in the ability of people to channel the flow of electrons. Later developments in electronics would expand the ways in which language played a role in both art and the technology itself. This book moves from the prehistory of the transistor to the transistor and beyond, following the capacity of electronics components to extend human sense perception. It is a story written by engineers and visualized by artists.

While the technology that the artwork interprets was often international in its origins, the role of the United States as a destination for émigrés from Europe and the rest of the world in the twentieth century, and the unique—and temporary—strengths that American industry enjoyed during its wartime and postwar boom years, bring this book's contents into focus at that particular time and place. A worldwide study of comparable phenomena—the vast legacy of postwar Swiss and Italian industrial graphic design, for instance—is better documented in books on those specific subjects.

In the background of this book are the stories of the founding electronics companies and their respective spheres of influence. Each of these companies, especially the largest half dozen, possesses a storied corporate culture and a compelling historical narrative. The four largest—General Electric, RCA, IBM, and AT&T—operated laboratories encompassing both pure research and invention, from which their business wings developed products. The books that focus exclusively on these companies have already been written. There are corporate histories, biographies of central figures, and critical counter-narratives that reveal an underside of waste, neglect, and abuse of people and of the environment on the part of the electronics industry. There are even books about these companies' own industrial design legacies. As the era of radio yielded to the era of computers, IBM in particular became a dominant sponsor of art about technology. Otherwise, the range of individual companies sponsoring art for industry generally mushroomed over time. The later chapters of this book, situated in the 1960s and early 1970s, express a broadened corporate and industrial landscape that had been transformed by thirty consecutive years of development and diversification.

Human beings only occasionally appear in artistic representations of emerging technologies. Stylized parts of the human body often appeared in artist-made works to symbolize the extension of the sensorium that electronics enabled, especially as those extensions approached the ability to mimic biological systems. But when real, whole human beings appeared, there was a striking consistency: women are at the center of the electronics manufacturing industries and are widely depicted throughout the decades, both photographically and in graphic art, as the makers, the crafters, the testers, and the researchers whose hands brought electronic components into being. At the dawn of the computer age, women were the first programmers and remained a strong presence in the programming field until the 1960s. I have chosen to include images of women at work, as they represent an important human historical context for electronics.

**NOTES ON THE SOURCE LITERATURE** *Fortune* covered many worlds for its readers, within which advances in electronics technologies, notwithstanding the July 1943 issue, were just one element among many. The rest of this book follows the evidence left not by *Fortune*, but by *Scientific American* and primarily by the vast pool of trade literature: magazines, catalogs, and pamphlets created to educate technical readers on subjects relating to the work that they do. The bulk of the images in *Inside the Machine* are from magazine advertisements published in a few key titles that were read by the electronics engineers, inventors, and workers of the era. *Business Week* and *Scientific American* were the more general of these; however, I spent far more time with two of the more technical titles, *Electronics* and the *Proceedings of the Institute of Radio Engineers* (known to engineers as *Proc IRE*; after 1961, the *Proceedings of the Institute of Electrical and Electronic Engineers*, or *Proc IEEE*). These periodicals include artwork on their advertising pages, but more importantly they support this book's exploration of the technical stories that inspired the visual source material. Both titles surprised and engaged me in ways I couldn't have predicted.

During its first four years, 1932–35, under the leadership of founding editor Orestes H. Caldwell, *Electronics* covered all developments relating to

Fig. i.11: SHAMELESS SELF-PROMOTION: AT THE CUSP OF THE SPACE AGE, THE INSTITUTE OF RADIO ENGINEERS MAGAZINE ADVERTISES ITSELF ON ITS OWN PAGES. THE ACCOMPANYING TEXT DESCRIBES THE SCOPE AND QUALITY OF THE READERSHIP. THE PROMINENCE OF THE APRIL 1960 SPACE ELECTRONICS ISSUE SPEAKS FOR ITSELF (*PROC IRE*, 1961).

the new technology seemingly without prejudice in regard to their potential applications. Electronic musical instruments, for instance, were reported on in *Electronics* as thoroughly as upcoming developments in vacuum tube technology. (Caldwell was cited by Congress for unduly favoring the interests of RCA during his career, but that's another story.)[10] In late 1935, Caldwell left the magazine to form the company Caldwell-Clements, which published two other titles consulted for this book: *Electronic Industries* and *Electronic Technician*. By the late 1930s, the interests of business had begun to elbow out the interests of culture and *Electronics* became less neutral. By the time its readership expanded in the early war years, it was more exclusively a journal about electronic technologies as they applied to business and industry.

The *Proceedings of the IRE* began in 1915 as the principal technical journal for electrical engineers. Opposite, in some ways, to the editorial trajectory of *Electronics,* the *Proc IRE* became increasingly broad in the scope of its articles over time. While the early volumes limited themselves to technical papers, by the 1940s engineers were using the journal as a forum for debate over issues within engineering. In the postwar era, the journal began to include nontechnical articles analyzing the applications of engineering to both biological and social environments. Beginning in 1960, with the publication of *IRE Transactions on Human Factors*, the magazine branched into specialized journals interpreting the societal impact of new technologies and other subtopics relating to electronics.

As this book traces the journey to the stars, it travels from tubes to transistors, from representation to symbolic language, and from abacuses to computers. Along the way it introduces artists whose works are mostly little known. Where possible, we'll catch a glimpse of the mechanics of mediation: How did the artists learn about the technology they were depicting? How close were they to industry? The answers to those questions fall across a spectrum within which the historical record is often thin. Where daylight does fall on the record of these relationships, it further illuminates the process by which artists reached for the electron, placed it on their palette, and painted it: the inside of the machine.

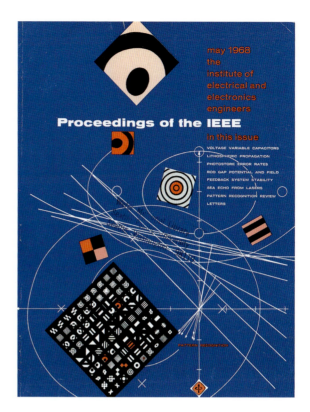

*Fig. i.12:*
A "PSEUDO-PSYCHEDELIC DESIGN . . . INSPIRED BY THE LINEAR CATEGORIZERS AND MASKS FOR EXTRACTING GEOMETRIC FEATURES DISCUSSED [WITH REGARDS TO] PATTERN RECOGNITION" (*PROCEEDINGS OF THE IEEE*, 1968).

## CHAPTER ONE

# THE ATOM, THE PLANET, AND THE TUBE

"Radio" is a way of thinking! Just as 'communication' needed to break its earthbound bonds of wire and take to the air, so industry is seeking and finding in radio controls new tools ranging from servo-mechanisms to electronic computers. This Is No Dream. . . . Just as radio physicists unleashed the 'radiation' power of the atom, so these same thinkers will harness it to industry. They have brought the picture of the world under your control by a knob in your home television—and have beaten the monotony of endless counting by the electronic computer.

—*Proc IRE*, advertising department copy, 1953[1]

**THE "EARTH BULB"** In 1938, a pair of unrelated events set the stage for a historic intersection of art and technology. That year General Electric started an experimental radio station to develop frequency-modulation mode (FM) broadcasting, a new and unproven technology at the time. The station, near GE's headquarters in Schenectady, New York, was a laboratory for the testing and development of the ultra-high-frequency vacuum tubes that would make FM radio broadcasting possible. The same year, Herbert Bayer, a prominent graphic artist and former teacher at the Bauhaus school, fled Europe for the United States. He brought with him a determination to build bridges between art and science.

FM radio technology had been invented in 1935 by Edwin Armstrong, a sometime researcher with the laboratories of RCA. In the late 1930s, the new invention's future was uncertain. RCA sought to thwart Armstrong's patents, believing that FM radio threatened the company's well-developed AM

Fig. 1.1:
HERBERT BAYER FOR GENERAL
ELECTRIC. THE "EARTH BULB"
(1942).

radio monopoly. Its corporate rival, General Electric (originally RCA's parent company), saw the potential of FM to challenge RCA's dominance in the radio market, and in 1938 it licensed the rights to FM technology from Armstrong.[2] In the fall of 1941, following three years of development, GE debuted three new "ultra-high-frequency triodes," the specialized vacuum tubes capable of powering FM radio. After it became clear that FM radio could be brought to the public,[3] GE commenced mass production of the tubes. They celebrated this breakthrough, and others, with the 1942 publication of a glossy promotional booklet, "Electronics: A New Science for a New World." The booklet was designed by Bayer, through an arrangement between GE and N. W. Ayer, the advertising agency that represented Bayer's commercial work.

The giant bulb on the booklet's cover (*fig. 1.1*) is the GL-880 transmitting tube, one of those that made FM radio possible.[4] The new triode also served the simultaneous emergence of broadcast television. The design details of the tube allowed for greater energy storage and control values, and therefore higher power, than had previously been available with smaller tubes.[5]

Clasped in the disembodied hands of humanity, Bayer's bulb is thrust into a sky in which atoms swirl like planets in the solar system. The tube forms a novel, glassine atmosphere holding planet Earth itself; Earth is both inside the machine and elevated within the solar system. The "Earth bulb" evokes a "new world," placing electronics at the heart of a designed picture of the modern age. Two powerful visual motifs operate: the vacuum tube itself, and, in the sky above, meshing with the planets, the familiar wide orbital loops of Rutherford's atomic model. The artwork combines these elements—in a gesture of tongue-in-cheek techno-heroicism—to convey the significance of a technology that works as a component part, not an end product. The solar *system* is the bulb's context, auguring the networked nature of the anticipated electronic future.

> Bayer's art is evidence that art and science are converging. Astronomy . . . engineering and art begin to have outward similarities, and to co-operate with one another. . . . The scientific attitude mingles with his creative sense. Stars, horizons, cube roots, architraves and plumb lines are among his models. Science today has marched out into space through sound, sight and self-projection, and this mastery over Earth's environment is reflected in much of Bayer's work.[6]

C. T. Coiner, an art director at N. W. Ayer who worked closely with Bayer, included this observation in his introduction to the catalog for a gallery show of Bayer's work the following year. Coiner had overseen production of the thirty-two-page General Electric booklet designed by Bayer, with full-color printing and original art on every page. The company's words, when combined with Bayer's ambitious graphic art, emphasized the promise for the civil world offered by technologies that were, at the time, most highly developed for wartime use.

A typewritten note attached to a copy of the GE booklet that was mailed out reads:

> Dear ———, Here is an advance copy of the publication "Electronics—A New Science for a New World." Its purpose is to assemble in one booklet a comprehensive story of electronics, its history, its important present, and something of its brilliant future, and to present that story in a way that reflects

the hope and promise of this youthful science. The publication is being sent to a broad list of executives in industry, and to educators, science teachers, and writers. Additional copies are available . . . [signed by W. R. G. Baker, an executive in General Electric's communications division][7]

In its task to place electronics at the center of readers' visualizations of the future, GE was fortunate to be working with Bayer. Born and raised in Austria, Bayer moved to Germany to study and then teach at the Bauhaus, at the time the world center of modern design. He was a noted designer of typography and graphic art in Europe in the 1920s and 1930s, but fled Germany after the Nazi Party condemned avant-garde art as "degenerate." He soon established a successful career in the U.S. that was exceptional in its ability to reach both fine art and commercial art audiences. Bayer's GE booklet predates his best-known work and the establishment of the Aspen Design Institute, which brought him lasting public fame; however, it anticipated many of the motifs that would characterize his later work. From the earliest days of his career, the Alpine-reared Bayer was focused on an integration between the natural environment and the modern, technological world.[8] His "Earth bulb" artwork and other illustrations in the GE booklet drew upon this history of creating graphic strategies to humanize technology by integrating it in an organic context. It would remain a dominant theme throughout his career, as in 1943 he constructed a model globe in the Museum of Modern Art,[9] in 1953 he created the *World Geo-Graphic Atlas* for the Container Corporation of America, and in 1955 he created a signature earthwork, a sculpture made from available ground materials, in Aspen, Colorado: the GrassMound.[10]

Bayer's General Electric pamphlet is an example of the reach of corporate communications. The pamphlet promotes applications of electronics in many industries, including medicine, agriculture, information technology, and industrial processes. However, radio was the big news of the day because of its immediate impact on everyday life. Radio led the public representation and understanding of

*Fig. 1.2:*
HERBERT BAYER FOR GENERAL ELECTRIC, ADAPTED [ALLIED RADIO CATALOG, 1947].

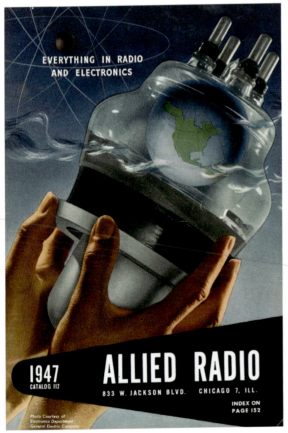

electronics, and Bayer's FM-oriented "Earth bulb" exemplified the trend. It was reprinted widely in GE magazine advertisements and on the cover of equipment catalogs as late as 1947. More significant for this book, Bayer's fame as an artist means that much more information about his work is available than is the case for most commercial artists. Bayer signed his work, and wrote about it and exhibited it. He inhabited the world of fine art with its freedoms as fully as the world of commercial art and actively sought to diminish distinctions between them, a legacy that is well documented in archives.

Although Herbert Bayer was hardly alone in his ability to identify as both a fine artist and a commercial artist, the traceability of his position was the exception rather than the rule. Often commercial artists were contractually prohibited from attaching their names to their work, and the paper trail leading to their identities has been lost. Since the postwar boom era, advertising agencies have undergone dramatic waves of consolidation, while many of the companies that engaged their services have been similarly consolidated or were lost to the forces of deindustrialization.

The problem of artist anonymity is not linked only to the artist's status. Significant artwork was often made to promote the work of small companies whose technological histories, and those of their corporate communications, were lost decades ago. General Electric, although it closed its Schenectady headquarters in the 1980s, placed its corporate archives with the science museum in that town, which maintains public access to them. The relationship between GE and Bayer, between artist and client, and artist and technology, sheds light on comparable processes that elsewhere are obscured.

**A SMALLER WORLD** The immediate context of GE's broad proclamations was the principal wartime application of radio: aviation communications. During World War I, radio technology was quite new and was used primarily as a method of communicating between ship and shore. Toward the end of that war, experiments with radio in aircraft were conducted, but radio did not yet affect the shape of air travel in a systematic way. As early as 1916, amateur radio operators (radio hams) assisted with Antarctic exploration, their skills creating a safety net that assisted the last great wave of terrestrial exploration of the interwar years.[11]

In 1927, Bell Telephone Laboratories, a GE competitor, purchased an aircraft and used it as a flying laboratory for refining air-to-ground and air-to-air radio communications.[12] Radio-assisted direction-finding beacons were developed to help pilots navigate in low-visibility environments.[13] The radio receivers of the day were large footlockers packed with vacuum tubes and wires, becoming progressively smaller throughout the experimentation of the 1930s. As large as they were, these radios had only 3 to 5 watts of power, extremely weak by today's standards. Their low wattage rendered them ill-suited for voice transmission, and early pilots communicated by Morse code.

By World War II, the airplane had become the centerpiece of radio-assisted geographic expansion. Radio made surface-to-air communications possible during the war and was used in that capacity all over the world. The sphere of human agency expanded during those years not just around the planet, as with terrestrial exploration, but skyward, to the altitude of flight. Our bubble of habitation expanded outward. Radio communications between ground and air even changed the relationship between space and time, as people could now speak to one another in real time around the curvature of the Earth, across time zones, even when at least one of those people was swiftly in motion. A whole new calculus of communication was born, and a new term was coined to refer to the particulars of flight electronics: avionics (*fig. 1.3*).

*Fig. 1.3:*
HAROLD FLUCKE FOR DELCO RADIO. THE INTEGRATION OF RADIO WITH AIRPLANES IN THE 1930S MADE AIRBORNE BATTLE OPERATIONS SAFER AND BETTER COORDINATED (*AERO-DIGEST*, 1945).

Delco Radio Products Mean Fine Performance

From compact auto radio sets to highly intricate radio and electronic equipment for the armed forces, Delco Radio products are distinguished by fine performance. Each unit reflects care and competence in engineering; each part represents advanced techniques in production accuracy. Tomorrow's civilian needs, like today's military demands, will benefit from Delco Radio's engineering vision, manufacturing precision. Delco Radio Division, General Motors Corporation, Kokomo, Indiana.

*Put Your Dollars in Action—BUY MORE WAR BONDS*

Delco Radio
DIVISION OF
GENERAL MOTORS

## THE EDISON EFFECT

World War II may have been the single greatest catalyst for the development of applications for vacuum tube technologies, but the war took place in the middle of the tube's technological development. The human ability to harness electricity progressed in fits and starts from early in recorded history to the late eighteenth century, when Benjamin Franklin began to learn how to harness and control its flow. During the nineteenth century, following Franklin's experiments, electricity was adapted to a number of technological processes, culminating in the work of Thomas Edison to create the very publicly visible electric "lamp," or lightbulb.

By the turn of the twentieth century, electric light was ubiquitous. The new "lamps" attained minor iconic status: glowing semi-spheres symbolizing the emergence of a well-lit age. The electric lightbulb was a new world, a sphere of possibility. The achievement of controlling electricity within a tube fueled the imagination of scientists and inventors. In 1884 Thomas Edison, following his invention of the lightbulb, patented a technique for drawing electrical energy out of a bulb through a wire.[14] This process became known as the Edison effect, although the extracted energy had to wait for subsequent inventors to be harnessed. In 1897, the English scientist J. J. Thomson identified the electron as the constituent element of the electrical charge that Edison's device manipulated. Subsequent research by Ernest Rutherford discovered the protons and neutrons that comprise the nucleus, and that together with a surrounding electron cloud form the anatomy of an atom.

The inventor Lee de Forest, for one, asked the question: How do you harness the power *within* electricity and make it do things beyond its illuminated spark—beyond the reach not in the sense of reaching outward, but reaching *inward* to the atom itself? That is the science of electronics, the process of controlling the flow of electrons, the constituent elements of an electrical charge. The *technology* of electronics is the application of this science toward useful purposes. De Forest, working in a Palo Alto, California, laboratory, drew upon Edison's research with a wafer structure inside the bulb that channeled electrons.[15] His working bulb, called an audion, compelled electrons to move in particular ways and, when outputted to a wire, to impart their energy to a network outside the bulb. The audion used electronic energy pulled through the wire to amplify sound. Later, more developed vacuum tube models would sort and amplify an electronic signal by passing it through wafers and vacuum space. In *fig. 1.5*, de Forest's audion is depicted nostalgically in a 1959 recruitment advertisement for Lockheed's space electronics systems division, invoking sixty years of electronics history at the threshold of the space age.

With de Forest's audion putting electrons to work, two new glassine spheres entered our culture, following along after the lightbulb: the vacuum tube and the tiny atom itself. The term "spheres" is particular: the atom and its internal structure was a new world for human imagination to inhabit. Within an atom, electrons surround the nucleus in a shape that, when abstracted in a drawing by Rutherford, was believed to resemble the motion of planets around the sun. The term "planetary model" was used in the 1910s and 1920s to describe the structure of the atom as it was being explored in light of dis-

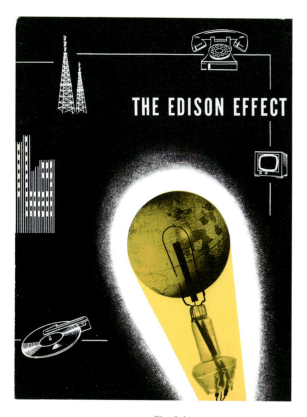

*Fig. 1.4:*
THE EARTH INSIDE A LIGHTBULB, FROM THE COVER OF AN EDUCATIONAL BOOKLET PUBLISHED IN 1951 BY THE THOMAS ALVA EDISON FOUNDATION. THE DESIGN, POSSIBLY INSPIRED BY BAYER'S "EARTH BULB," ASSOCIATES EDISON'S INVENTION WITH THE LEGACY OF THE ELECTRONIC AGE.

*Fig. 1.5:*
LOCKHEED. AN AUDION STANDS
AT THE GATEWAY TO SPACE
(*ELECTRONICS*, 1959).

*Fig. 1.6:*
AL SCOTT FOR HOOVER
INDUSTRIES, 1946.

coveries by Rutherford and subsequent refinements to Rutherford's model by Niels Bohr. Rutherford was a contemporary of Einstein's, at a time when Einstein's discoveries were opening the door to understanding the relationship between space and time. Even as these and other physicists added significant refinements to science's understanding of the electron, the early shorthand of "planetary model" stuck. The link between the two became more widely articulated as the Rutherford–Bohr model was explored, and reinforced by significant new astronomical investigations into the structure of the solar system in the first quarter of the century.

After a lull in commercial artwork during the Depression, the motifs of planet, solar system, atom, and vacuum tube found new applications in the wartime boom years. The vacuum tube, with its rounded, organic shape, became a new visual container for imaginations about technology. The *Fortune* article described in the introduction includes the internal headline "The World Inside a Tube," and Bayer's "Earth bulb" was the earliest and most influential exemplar of this design trend. In 1946, an illustrator named Al Scott set aside the tube itself in favor of a direct expression of the link between the atom and the planet (*fig. 1.6*). His artwork decorated the cover of an electronics textbook

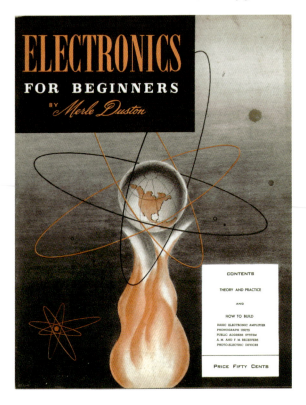

by Merle Duston in which the author declared, "This is a scientific age. This is the atomic and electronic age . . . The developments of World War II prove that we are on the threshold of many great new developments which will affect all of our lives."[16]

Bayer's "Earth bulb" appears on the GE booklet in a glamorized condition, bare of its cumbersome water jacket, the oversized wraparound device that was needed to keep the big, hot tube cooled to operational range. A more practical and decidedly less glamorous-looking version appears on the cover of Concord Radio's 1952 equipment catalog. A similarly disembodied, and authoritative, hand holds the bulb, and the bulb, along with other components, floats in the firmament. But the bulky jacket somewhat interrupts the bulb's authority to reframe a new world.

While the GE booklet was aimed at decision-makers, subsequent quotations and imitations of its motifs in trade literature brought its message to a working-class audience. The dramatic shift in scale that the advent of electronic technology represents remains embedded in these images. At least as much as the GE booklet, trade publications were instrumental in communicating the developing belief that new technologies—electronic technologies specifically—would transform everyday life and eventually bring the skies, even the solar system, closer to our world. Many postwar radio repairmen, such as those reading the Concord catalog, would have been returning soldiers.[17] They would therefore have been differently prepared—perhaps far more ready than Stateside industrialists—to receive promotional claims about the symbolic significance of electronics. They would have had firsthand experience of how radio communications had changed the war, including the impact of Germany's V-2 rocket.

Imported from Germany and guided by electronics, the V-2 would eventually bring space within our reach in the 1950s. But in the late 1940s and early 1950s, the space age was yet to come, and its ideas were still one part fantasy, one part ideology. Only with the developments in electronics of the 1950s would the graphic alignment of the atom and the planet be proven in the real world through satellite technology. In the meantime, the motion of electrons around the nucleus of an atom remained a peaceful visual synecdoche for the solar system.

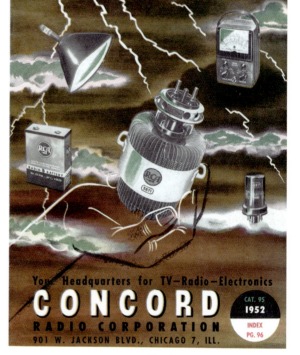

**A DIMENSION OF SOUND** After de Forest's audion, the development of the vacuum tube was distributed across radio workshops around the eastern U.S. Of all the human senses, the one to which the electronic vacuum tube's applications were most relevant is hearing. Vacuum tubes enabled radio, sonar, and beacon-assisted flight—all technologies of the audiosphere. With the advent of FM radio, the global expansion made possible by aircraft-based communications was duplicated in living rooms, and audio technologies multiplied around the world. In the U.S., a postwar surplus of industrial capacity would bring FM radio to the public, widening the worlds of those at home.

In 1944, two years after Bayer's pamphlet, General Electric continued its grand graphic promotions of its tube line with additional pamphlets designed by less well-known artists. In *fig. 1.8*, Bayer's style in blue is lightly imitated, and the Rutherford model of the atom shares the stage equally with the tube itself.

*Fig. 1.7:*
THE TUBE IN THIS ARTWORK IS VERY SIMILAR TO THE TUBE IN FIG. 1.1 (IT'S THE RCA VERSION); THE DIFFERENCE IN APPEARANCE IS CAUSED BY THE PRESENCE OF THIS TUBE'S COOLING WATER JACKET (CONCORD RADIO CATALOG, 1952).

The tube itself was a marvel of containment. Its form echoed that of paperweights and snow globes, and its fragile constitution made it both a novelty and a delicacy. And yet, its role as a transformative technology demanded that its interior be explored as a possibility space of real importance. During the late 1940s, vacuum tubes were combined with electromechanical calculators, switching systems, and data processing machines, among other electronic elements, to form early computers, a process explored more deeply in chapter 6. As vacuum tubes became identified with computing as well as with radio, their identification with the sense of sound was expanded to include an identification with the human mind.

The alignment of tubes with the realm of thought harmonized with the existing graphic motif of the sky. In IBM's 1948 advertisement (*fig. 1.9*), a nocturnal sky offers a visual field against which the vacuum tube is shown off nicely. In this case, the artist is aligning a stylized cloudscape with the realm of

*Fig. 1.8:*
GENERAL ELECTRIC, 1944.

INSIDE THE MACHINE **38**

computing, placing a suitably philosophical Grecian head where a few years earlier, in Bayer's work, there had been a planet. The IBM design department was formed the same year this ad was published and by the 1960s would be an international leader in the use of art both for industrial purposes and as a tool for public education about science.[18] This ad was most likely produced by an advertising agency, as the early years of the company's internal design department were focused on the industrial design of computing hardware. Nonetheless, this image was doubtless created through a collaboration between IBM and a graphic artist, and as such presages the company's close engagement with combining art and science.

In keeping with the tube's morphology of containment, the most advanced technological use of tubes was itself contained, in secrecy, during its earliest years. Computer technology was largely developed in closed laboratories, its military origins demanding from the very first that its development fall under the watchful dual eyes of government and research-based science, most often academic science. The application of tube technology to existing mechanical and electromechanical adding machines (those aided by electricity) was a process that developed in many stages, in many places, with contributions by many inventors, becoming fully electronic during the 1940s. Vacuum tubes formed the memory units of the earliest mainframe computers; bulbs in early computers stored information as components of binary circuits, enabling the retention of data and the process of computation. To an even greater extent than radio, computing technology was an offspring of the war effort. The German engineer Konrad Zuse built the first electronically assisted computational machine in 1941, though it was little known outside Germany until much later. The British code-breaking Colossus computers were the most powerful computers of their era, but they were kept tightly secret until after the war. In the United States, the most singular developmental landmark was the ENIAC (the Electronic Numerical Integrator and Computer), developed by J. Presper Eckert and John Mauchly

*Fig. 1.9:*
IBM. PROMOTION FOR PUNCHED-CARD PROGRAMMED ELECTRONIC CALCULATORS AND OTHER DEVICES. THE WORD "COMPUTER" DOES NOT YET APPEAR (*SCIENTIFIC AMERICAN*, 1949).

with teams of associates and assistants. The machine first worked well in 1945, but it was not until 1950 that the first stored-program electronic computer (the UNIVAC, Eckert and Mauchly's civil project) was manufactured for commercial consumption. Prior to UNIVAC, virtually all computers had been purpose-built for military applications.

The early history of computing can only be understood as a wide cluster of roughly simultaneous developments during the late 1930s and 1940s, taking place in both Europe and the United States, all leading toward the same general outcome of vacuum tube-powered electronic computing. Common to all these efforts was the harnessing of electronic power to assist processes of computing that had previously been conducted by hand, or by mechanical or electromechanical machine. The vacuum tube boosted these processes so phenomenally that the tube-driven electronic computer constituted a qualitative leap beyond the capacity of electromechanical computing. During this flurry of activity, many of the largest commercial electronics laboratories had dedicated labs for computer development. It was in this environment that IBM and Bell Laboratories ran their early computing experiments, putting their achievements on a par with the grand experiments conducted by the combined forces of academia and the military. The University of Pennsylvania

*Fig. 1.10:*
REMINGTON RAND, C. 1951. REMINGTON RAND ACQUIRED THE ENGINEERS J. PRESPER ECKERT AND JOHN MAUCHLY AND THEIR UNIVAC COMPUTER IN 1950 AND BEGAN TO PROMOTE THE MACHINE TO INDUSTRY SOON AFTER.

hosted the ENIAC; Harvard University hosted the Mark series; and Princeton University hosted the research into computing architecture conducted by John von Neumann at the Institute for Advanced Study. All these early efforts yielded room-sized, tube-driven prototypes of analytic supercomputers. The commercial computer industry of the 1950s was built out of the entire resulting particulate cloud of knowledge.[19]

Graphic motifs associating the computer with the human mind continued to develop for the field of artificial intelligence, a topic addressed in later chapters. In the literature of the business community, however, the artwork yielded to more deskbound applications of information technology. In *fig. 1.10*, a pamphlet cover promotes Remington Rand's acquisition of the UNIVAC project in 1950, the first general-purpose computer manufactured for commercial use. Remington Rand had been an office equipment company since the nineteenth century, making everything from typewriters to desktop calculators. With its acquisition of Eckert and Mauchly and their work, the company catapulted itself into the ring with the other business-machine-to-computer giants, led by IBM.[20] The surrealist pamphlet artwork references both the safe enclosure of a snow globe and the unleashed capacity of office work. Freedom and containment coincide: a tiny working office sped up by technology yet enclosed within a tube.

The pattern of intensive laboratory development during the 1940s followed by a "market" decade in the 1950s repeats itself across the histories of many electronic components and devices. Cathode-ray tubes and transistors, for example, share this rough chronology with vacuum tube-driven computers and FM radio. The pattern is explained by an intense wartime investment in war-related infrastructure and basic research science, followed by a resulting postwar surplus of industrial and manufacturing capacity that was well suited to converting war-related technologies to civil uses. With the U.S. nearly alone among industrialized countries in the 1950s in possessing an undamaged manufacturing base, its industry was particularly, if temporarily, well-situated to dominate domestic and world markets for new technologies.

## THE COLD WAR RUTHERFORD-BOHR MODEL

In the 1950s, as the Cold War matured, the cultural signification of the Rutherford-Bohr model shifted. Many today may associate those iconic loops with grim radiation warnings left over from the Cold War, but the symbol had other, earlier uses. As the illustrations in this chapter have shown, it signified technologies as benign as radio and radar, and was even used to offer an optimistic image of the future of technology.

The changing meaning of this symbol—and its simultaneous, divergent meanings—offers an indication of the extremely rapid pace of technological change in the mid-twentieth century. Elsewhere in society, by 1950 variations of the Bohr-Rutherford model symbolized the culture of danger and risk surrounding the new and shocking technology of atom-splitting for purposes of war. A simplified and standardized version of the atomic symbol appeared in countless government documents, on industrial signage, and in popular literature. Sometimes it specifically indicated potential danger from radiation; in other instances, it became a shorthand for the malaise that had seeped into society from the shadow cast by atomic weaponry. Ultimately, its design-friendly loops and simplicity rendered it an essential component of popular modernism. Stylized Rutherford-Bohr model designs and diagrams decorated mid-century Modern domestic environments, to the point that a magazine of twenty-first-century nostalgia for this design era is called *Atomic Ranch*.

Between 1940 and 1960, three separate communications campaigns were conducted for public understanding of the atom, each of which used the model to convey very different meanings. Following the detonation of the first atomic bombs, atom-splitting technology was developed into an energy source. Nuclear fission and the process of isotope decay were developed to power everything from large generating plants to small atomic batteries in civil spacecraft. The campaign for acceptance waged by the civil nuclear industry, the industry of atomic energy, was also symbolized by the Bohr-Rutherford model. This campaign included advertisements in *Nucleonics* magazine, and more mainstream magazines such as *Business Week*. *Nucleonics* was founded by one of the editors of *Electronics* in a gamble that nuclear energy would be the next big thing after electronics.

The electronics industry was able to successfully use the model to signify its work because electronics had beat the atom-splitting technologies to the punch. The relatively nondestructive process of harnessing the motion of electrons to do work turned out to be the tortoise that won the race—at least, this race for an association with the popular symbol; as late as the mid-1960s, the familiar multi-loop icon was at work promoting electronics, in visual terms that were clearly unrelated to the iconography of the nuclear threat. By

*Fig. 1.11:*
BANKER'S TRUST. VISIBLE LANGUAGE: HERE A PROFUSION OF DIGITS IS A GRAPHIC MOTIF OF ITS OWN IN A GALAXY SWIRL CONNECTING A SKY-BORNE VACUUM TUBE TO THE DESKS OF DOZENS OF WORKERS (*BUSINESS WEEK*, 1950).

"Vision is Indispensable to Progress"

Electronic tubes that lift loads of laborious work from office desks!

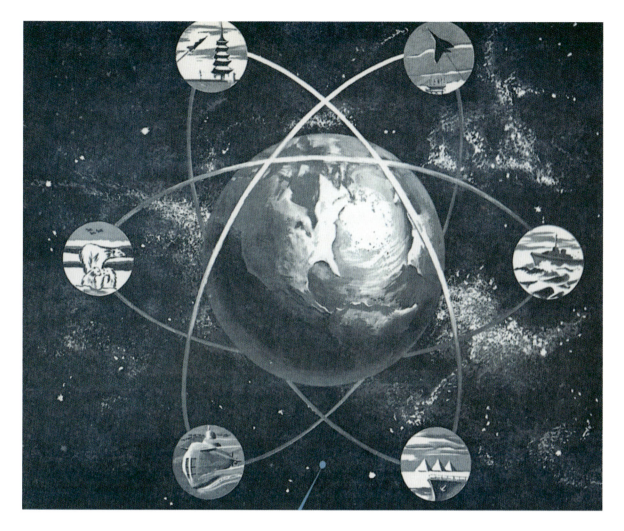

*Fig. 1.12:*
CONTINENTAL ELECTRONICS.
THE ATOM AND THE PLANET
(*PROC IRE*, 1961).

1959, however, there was a bit of pushback. In its critical "Copy Chasers" column of September 1959, *Industrial Marketing* called for advertisers to "help stamp out ad clichés," fingering "the atomic cliché" for having been diluted to promoting everything from farm machinery to cardboard boxes.[21]

Vacuum tubes faced obsolescence in mainstream applications in the early 1950s, as we will see, though they have remained active in some industrial applications. In particular, they are still valued by audio connoisseurs, as nothing amplifies sound better than a nice fat tube. But after the invention of the transistor, vacuum tubes ceased to symbolize the future. Only decades later did they return to the periphery of the public eye as symbols for forgotten qualities of the pre-transistor era.

## CHAPTER TWO

# TUBES THAT SEE: CATHODE-RAY TUBES

Demobilization day will find television a fully explored but wholly unexploited field . . . I think it quite likely that during the postwar period television will be one of the first industries arising to serve as a cushion against unemployment.

—James Lawrence Fly, Chairman, Federal
Communications Commission, 1942[1]

**OSCILLOSCOPES: THE FIRST CRT DEVICES** The undulating zigzag of today's heartbeat monitor is a visual motif that has been with us for well over a century. All modern screens, from television and video game monitors to pocket devices and airport security monitors, are descended from the first cathode-ray tubes (CRTs). The history of the CRT is the history of electronic visuality. These specialized vacuum tubes convert an electronic signal into an image that is projected onto the front of the tube: the staccato movements of points on the screen of an oscilloscope, an early cathode-ray tube device, visualize whatever changes, or oscillations, take place over time in the signal that is being monitored. CRTs were the first electronic screens, and also the first electronic *eyes*. They opened a channel of vision onto the moving flow of electrical current, making visible a phenomenon previously unseen. The incoming signal was previously accessible only as sound. The "ping" of radar and the squeals and whistles of an old-time radio being tuned are soundscapes made visible by CRT technology.

Cathode-ray tubes are long-lived technological links between nineteenth-century vacuum tubes and contemporary signal-monitoring technologies, including LED and LCD computer monitors. The basic architecture

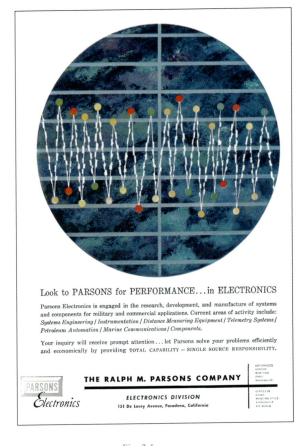

*Fig. 2.1:*
THE RALPH M. PARSONS
COMPANY (*SCIENTIFIC
AMERICAN*, 1961).

of a CRT is a vacuum tube with an expanded outward face, the screen, coated with a fluorescent or reflective material that can make an electron projection visible to someone looking at it from the outside. Inside the tube, an electron gun and a series of focusing and deflecting plates shape the flow of electrons into a "picture" that becomes visible when projected on the outward face. The earliest, simplest cathode-ray tubes projected the green fluorescent trace of an undulating signal, displaying its variance by moving across the otherwise dark screen: the oscilloscope.

Oscilloscopes display data prompted by fluctuations in the flow of electrons, fluctuations that can signify changes in the status of a system, either living or assembled. One of their earliest applications was to modulate electrical current that was applied to the human body for therapeutic purposes, presaging the era of medical electronics.[2] The oscilloscope became an extension of the human senses. Early oscilloscopes, like the earliest television screens, were small, more like an "eye" that drew the viewer to "eye contact." Over time, the development of the computer monitor turned that "eye" into a "face," the face of the new companion that computers would eventually become. In the early twentieth century, they were among the earliest devices that began to socialize people to looking at electronic screens, a notable turning point in the relationship between technology and the human body.[3] Television later socialized people to screen-viewing on a mass basis.

The oscilloscope was quickly incorporated into many phases of industrial research and testing. By the first quarter of the twentieth century, oscilloscopes were used to monitor signals from measurement tools, signals indicating voltage, current, and waveforms. Mechanical devices could be tested for vibration, balance, and speed with an oscilloscope that mapped an incoming signal onto a grid that finely measured the variations in the signal. The familiar zigzag shape of a signal tracing across the face of an oscilloscope was the first live screen image that people routinely looked at.

In converting an electrical signal to a fluctuating visual trace, oscilloscopes enabled an unprecedented level of precision in measurement. Oscilloscopes let people literally see variations in the behavior of a system at a refined level of granularity. In that sense, oscilloscopes not only extended human vision, they also expanded our reach across different scales of measurement. Degrees of variation previously beyond the human ability

Fig. 2.2:

TUNG-SOL ELECTRIC INC. FACE TO FACE: IT TOOK SEVERAL DECADES FOR THE EXPERIENCE OF GAZING AT A MONITOR TO BECOME A COMMON ONE. HERE, AN OSCILLOSCOPE IS USED TO TEST VACUUM-TUBE CALIBRATION (*ELECTRONICS*, 1960).

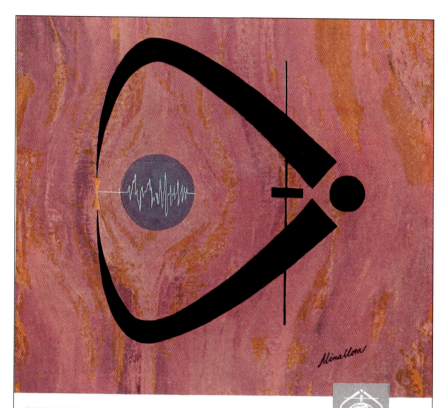

## MEASURING BY TATNALL

Tatnall Measuring Systems Company is a subsidiary of The Budd Company. Its function is to invent and produce modern physical testing equipment to meet today's insistent demands, and to anticipate the increasingly complex requirements of industrial and technological progress.

It would be difficult to exaggerate the dependence industry must place on measuring systems. In less than a generation precision has spanned a gulf greater than the difference between the dollar watch and the chronometer.

Industry needs to know—and know exactly—the influence of temperature, incidence of fatigue, effects of vibration, reaction to tension, torsion, compression and stress, on materials and structures of all kinds in any environment, under both static and dynamic conditions.

Tatnall has made important contributions to the means of measuring. But the needs that have been met are as nothing compared to those to come. The Budd Company, Philadelphia 32.

**INDUSTRY IN MOTION**

The challenge of change is being met by many new Budd projects in the fields of nuclear energy, supersonic flight, radiography and plastics. These, of course, in addition to our established place in the automotive and railway passenger car industries.

Fig. 2.3:

THE BUDD COMPANY. THE USE OF OSCILLATORS AS QUALITY CONTROL TOOLS, PRECISION MEASUREMENT TOOLS, AND INDUSTRIAL PROCESS ACCESSORIES MUSHROOMED IN MID-CENTURY, AS DID THE USE OF OTHER KINDS OF VACUUM TUBES. RAUL MINA MORA'S PAINTING FOR BUDD INTERPRETS THE USE OF AN OSCILLOSCOPE AS A TOOL FOR PRECISION MEASUREMENT (*BUSINESS WEEK*, 1958).

Fig. 2.4:
JET PROPULSION
LABORATORY. A RADAR ARRAY
PROJECTS ELECTROMAGNETIC
WAVES INTO THE SOLAR
SYSTEM (*PROC IRE*, 1958).

## PIONEERS IN EARTH-SPACE COMMUNICATIONS

The exploration of outer space will take a new step forward with the completion of the new giant radio antenna being installed by JPL near Barstow, California. This huge "dish," 85 ft. in diameter, will enable the Laboratory scientists to probe still farther into space problems.

Information thus obtained and combined with lessons still being learned from the successful Army "Explorer" satellites, will provide invaluable basic data for the development of communication systems to serve space exploration programs. Long range communication will begin as a one-way link from space to earth, developing later into tracking and communicating with lunar vehicles at far greater ranges.

This activity will be part of a great research and development program to be operated jointly by JPL and the United States Army Missile Command.

**jpl**

### JET PROPULSION LABORATORY
#### A DIVISION OF CALIFORNIA INSTITUTE OF TECHNOLOGY
#### PASADENA · CALIFORNIA

OPPORTUNITIES NOW OPEN ▶ APPLIED MATHEMATICIANS · ENGINEERING PHYSICISTS · COMPUTER ANALYSTS · IBM-704 PROGRAMMERS
IN THESE CLASSIFICATIONS ▶ FIELD ELECTRONIC ENGINEERS · SENIOR R.F. DESIGN ENGINEERS · STRUCTURES AND DEVELOPMENT ENGINEERS

to measure suddenly became visible, and thereby subject to manipulation. With the visual feedback from oscilloscopes, people could reach inside their machines and reengineer their internal processes, an advantage that greatly enhanced twentieth-century industrialization.

**RADAR** The primary wartime application of CRT-oriented technology was the use of radar screens. Radar (Radio Detection And Ranging), was developed nearly simultaneously by, and through collaboration between, American and English scientists, in the 1920s.[4] Its development was based on insights that originated as early as 1886, when the German physicist Heinrich Hertz demonstrated that radio waves could be bounced off solid objects. Radar, in essence, uses reflected radio waves to "see" distant objects. The signal images returned by radar arrays are captured and displayed on cathode-ray tube screens.[5] It was developed rapidly in anticipation of World War II, which it then profoundly affected by enabling sightless "vision" of ships, submarines, and airplanes.

After the war, in 1946, members of the Army Signal Corps bounced radar waves off the moon, a trick now performed by radio hams at local

*Fig. 2.5:*

RCA. A NEWLY IMPROVED RADAR MONITOR IS ABSTRACTED WITH A MODERNIST TURN. CRTS THAT ARE SPECIALLY ADAPTED TO DISPLAY RADAR SIGNALS ARE KNOWN AS PLAN POSITION INDICATORS, OR PPIS (*ELECTRONICS*, 1959).

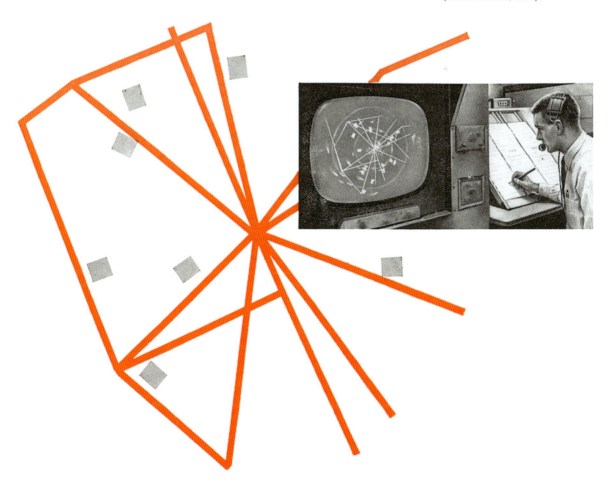

science fairs around the world. At the time it was a transformative leap, opening the door to radar-based astronomy and geophysical sciences. Ultimately, the sightless detection of distant quasars, pulsars, and other stellar phenomena through radio astronomy tremendously expanded the reach of astronomical science.[6]

The inherent visuality of CRT displays opens questions about the displays' relationship to graphic art. The process of converting an incoming electromagnetic signal into a graphic display yields an abstracted image. These images were early methods of graphic data visualization. As abstractions, they tended to conform to basic geometric shapes and patterns: perhaps angular lines, or waves made symmetrical by the mathematical logic of sine and cosine. The undulating lines and waves became graphic motifs for electronic data that persist to this day. In other kinds of visualizations, such as radar, early monitors were round, and the radial sweep of the scanning beam cut a geometrically perfect radius across the circle. Later monitors, such as the one shown in *fig. 2.5*, could conform to the shape of television screens. On any kind, objects being monitored show up on the screen as points, while motion scans show up as angular lines. Given this relationship of the technology to geometric form, it is no surprise that artwork made to represent it drew strongly on those geometries. In *fig. 2.5*, RCA promotes its new air traffic monitoring screen, with a small photo inset to reveal the basis for the abstract-modern line art that dominates the page. Elsewhere, monitors demanded, through their novelty, some level of realism in their depiction in order to convey what they were.

## 1946: CATHODE-RAY TUBES AND TELEVISION

All CRTs visualize a signal that changes from one moment to the next. Over the course of the twentieth century, cathode-ray tubes were developed and differentiated many times over, with specialized CRTs finding major applications throughout industry, research science, medicine, and television. Now culturally dominant, television was slow to develop after its invention in the first quarter of the century. By 1928 mechanical TV (a signal transmission process that paired mechanical image scanning techniques with electronic components) was broadcasting in a few places, but fully electronic television systems took longer. As is the case with a great number of technological developments, electronic television was invented nearly simultaneously by different people in different places.

Two of the lead inventors were Philo T. Farnsworth and Vladimir Zworykin; one was from a Utah Mormon farming family, while the other was an immigrant engineer from Russia. Farnsworth's early television camera tube relied on image dissection, a process similar to photography in which a photosensitive surface is exposed to the scene to be televised.[7] Zworykin, an RCA staff inventor, developed the iconoscope, a television camera tube with more features in common with basic CRT mechanics. All television tubes (other than the Farnsworth image dissector) are based on the electron distributor, a component within the CRT that "paints" electron patterns onto the outward-facing surface of the tube. The "painting" of electrons gradually encompassed the use of color, leading to color television. The names of different television camera tubes and receiver tubes vary by manufacturer: by the 1940s most receiver tubes were called kinescopes; their partner camera tubes were first called iconoscopes, then orthoscopes.

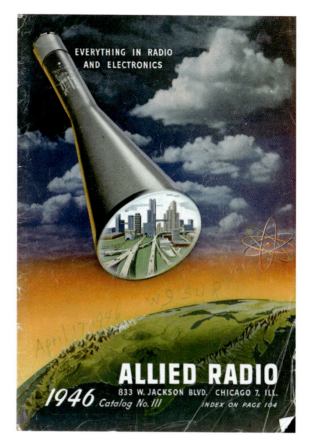

*Fig. 2.6:*
ALLIED RADIO CATALOG, 1946.

The artist whose visualization in *fig. 2.6* placed the kinescope in the firmament of cultural memory is unknown. It's hard to compare this 1946 Allied Radio catalog cover, aimed at hobbyists and repairmen, with the impact of Herbert Bayer's higher-profile work on behalf of the GE triode, but the distribution of Allied Radio catalogs was probably in the tens of thousands at the time. On the catalog cover, a gargantuan kinescope tube hovers over the eastern half of the United States; its size relative to the planet makes it appear to be about 5,000 miles long. The compositional elements of Bayer's "Earth bulb" are all present here: the planet, the tube, and the Rutherford atomic symbol, not to mention the sky and a vividly chromatic lower atmosphere layer. This cover art imbues the CRT with a sense of futurism, as the image on its face depicts a dense urban environment, thick with skyscrapers and freeways (the future), while the landscape below is distinctly verdant and apparently untouched, even borderless (the past). The U.S. interstate highway system had not yet been built, so the city in the tube would have appeared decidedly futuristic to the eyes of 1946. Even more noteworthy is the hovering "eye" over North America. Looked at today, it's hard to not to form an association between this image and our contemporary surveillance culture.

Television emerged and developed in cycles of laboratory trials—and

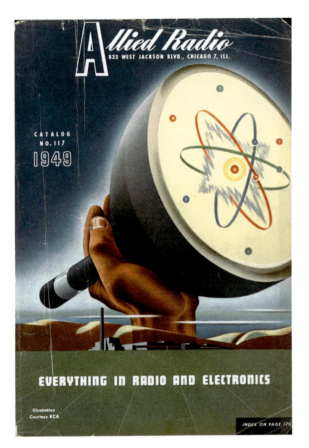

court trial as well, as there was considerable contestation over the technology in its early decades. RCA badly wanted Farnsworth's image-dissector tube for its television system, but fought the inventor for many years over the rights to it. After enormous expenditures of time and resources, in 1939 Farnsworth succeeded in securing from the company—in that era a corporate colossus—its first ever patent licensing agreement.[8] The Great Depression also thwarted the development of television broadcast systems, as the costs of developing and building such systems, and the projected cost of consumer receiving sets, were quite high. There was also fear within the electronics industry that television would "kill radio," especially if it were implemented unevenly.[9]

Some degree of realism was required in the artwork that interpreted CRTs, so that people would understand what they were seeing. Yet with CRTs, that realism also morphed into surrealism, in an expression, or anticipation, of the technology's massive impact. In surrealism, differences in scale within a frame defamiliarize objects that are otherwise depicted in quotidian realism, with an objective of pointing to hidden truths. The giant tube floats above Earth (*fig. 2.6*); the giant fist rises from behind a factory (*fig. 2.7*); Lilliputian people workshop the device into being (*fig. 2.8*). Why, in these images, is the human hand, and the new "eye," or "face" (the tube), so dramatically disembodied and resized? Perhaps because vision is so central to our lived experience. Our eyes are the sense most strongly connected to our awareness. Our other senses can be blocked, but if we can see then at least we know what is happening around us.

New and shorter big screen 16-inch kinescope developed by RCA scientists.

## Problem: shrink the television tube, but keep the picture **big!**

Some rooms accommodate grand pianos, a small spinet is right for others. Until *recently*, much the same rule held true for television receivers. Your choice was governed by room space.

Now the space problem has been whipped by RCA scientists, who have *shortened the length of 16-inch television "picture tubes" more than 20%!* All the complex inner works —such as the sensitive electron gun that "paints" pictures on the screen—have been redesigned to operate at shorter focus, wider angle. Even a new type of faceplate glass,

Filterglass, has been developed for RCA's 16-inch picture tubes—on principles first investigated for television by RCA.

Filterglass, incorporating a light-absorbing material, improves picture quality by cutting down reflected room light . . . and by reducing reflections inside the glass faceplate of the kinescope tube itself. Result: richer, deeper black areas and greater contrast in the television picture!

*See the newest advances in radio, television, and electronics at RCA Exhibition Hall, 36 W. 49th St., New York. Admission is free. Radio Corporation of America, Radio City, N. Y.*

New RCA Victor home television receiver, with big 16-inch screen— now *more than 20% shorter* in depth.

## RADIO CORPORATION of AMERICA
### World Leader in Radio — First in Television

1931—Pioneer Du Mont
cathode-ray picture tube.

# A new gift of VISION for modern man...

First comes vision, then development, then fulfillment. From the fountains of Du Mont vision flow achievements to enrich the world.

Dr. Allen B. Du Mont foresaw the great future of television when he perfected his pioneer cathode-ray tube in 1931.

From that tube he developed the priceless gift of the Du Mont Laboratories to mankind . . . all-electronic television, presented to the public in 1938.

Today modern television, pioneered by Du Mont, brings the richest daily entertainment in history within the reach and means of almost every home. And wherever television goes, the people receive new vision too, for better living and fuller citizenship.

Now again, Du Mont Television looks beyond past successes . . . producing, this year, the first practical large-screen color television tubes, to advance the whole industry's progress. It is through vision, development, and achievement, such as this, that Du Mont continues to be "first with the finest in television" . . . as in all telectronics for the home, industry, science and national defense!

*Write for free 40-page booklet, "THE STORY OF TELEVISION",
Allen B. Du Mont Laboratories, Inc.
Executive Offices, 750 Bloomfield Ave., Clifton, N. J.*

1938—Pioneer Du Mont
television receiver.

1954—Modern Du Mont Teleset.*

VISION IS THE **DU MONT**® DIMENSION

First with the Finest in Television

The phrase "the persistence of vision" was borrowed in 1978 by the science fiction author John Varley as the title of his signature work, so to many it suggests the mind-bending nature of visionary sci-fi. However it technically refers to the relationship between the human eye and the human mind that results in our ability to see continuously, even when the image we are looking at is flickering. The human eye retains a "cached" image for one-tenth to one-fifteenth of a second, allowing us to perceive a sequence of flashing images as a continuous moving picture. Television images on the front of a kinescope tube are formed by patterns of electrons flash-projected onto the front of the screen by a pulsing cathode-driven electron gun.

The electronic expansion of that sense changed our very sense of humanity, and would ultimately push us toward our contemporary investigations into post-humanity—the state in which human life has been fundamentally altered by, and adapted to, technology. These images and the technologies they represent contributed to the total impact of electronics over its first hundred years: the expansion of the human sensorium and the resulting decentralization of the human body into networks of sense-extending devices. Astronomers make our world larger by virtue of what they can see. They still rely on telescopes, but today their firsthand experience is often of CRT monitors that translate for daytime eyes images that the scopes have gathered overnight. The advent of television—and all CRT-enabled distance vision devices—expanded our world beyond the normal scale of sensory expansion.

The remarks by the FCC chairman that open this chapter express that in 1946 television was expected to be the next big postwar domestic technology. Expectations about television unfolded unevenly at first. The textbook *Electronics for Beginners* (*fig. 1.6*), for example, also published in 1946, gives television only a glancing mention. Yet the Allied Radio catalog cover offers as singularly visionary an image of the kinescope as there has ever been. The tube's cultural impact was ultimately immense, in keeping with this image. It made possible nothing less than the phenomenon of looking at other people without oneself being seen.

In 1954 the DuMont Laboratories of New Jersey, which also manufactured televisions, were an underdog TV network. DuMont made a significant contribution to the development of network programming before its efforts were eclipsed by the emergence of the major twentieth-century networks.[10] The company produced the graphic ad (*fig. 2.9*) illustrating four stages in the historical development of television: at top left is the cathode-ray tube; in the middle, a 1938 television; and at lower right is a "modern," 1954-model large-screen television. Behind the three, the iconic jagged zigzags of the electro-

(*opposite*) *Fig. 2.9:* DUMONT. A THREE-STEP PROGRESSION FROM CRT TO "OLD-FASHIONED" TELEVISION TO "MODERN" TELEVISION, COMPLETE WITH BASEBALL COVERAGE (*BUSINESS WEEK*, 1954).

magnetic spectrum illustrate AM and FM carrier waves. In 1954 television was just beginning its incursion into everyday life. For most people, even those with television, the device's means and origins were not necessarily familiar. This ad artwork explains much in a few strokes of the brush.

## CATHODE-RAY TUBES AS ARTISTIC MEDIA

The display of changing visual information was their central purpose; as such, cathode-ray tubes expanded *time* as a new dimension of electronic component capacity. Encompassing both time and visuality, CRTs have a more complex relationship to art, and art-making, than did vacuum tubes. The history of the oscilloscope includes the earliest use of electronic components themselves as artistic media. Cathode-ray tube interfaces, according to the media scholars Kevin Hamilton and Ned O'Gorman, "depend on sensors that monitor conditions and report change, graphically, via sampled moments of time . . . the viewer experiences this succession of sampled moments as a real-time window onto remote phenomena."[11] (The contemporary technical term encompassing screen-based art, film, and digital media is "time-based media," to differentiate these from other art forms.)[12] In this sense, all CRTs, starting with the humble oscilloscope, anticipate live video. While lightbulbs furthered the development of motion picture cameras and projectors, making possible one very well-known new art form (cinema), CRTs are perhaps the first electronic components that were themselves an artistic medium. As early as the 1930s, early "video" artists were playing with signal forms on oscillator screens.

The artforms to which CRTs lent themselves spanned both visual art and music. To the sense of sound, the theremin is an oscilloscope that is also a musical instrument, one that "plays" an electron flow, rendering its fluctuations as changing musical sounds. Its unique electronic vibrato sounds like the tuning of a measurement oscillator that has encountered some unusual spikes in its data survey. "[T]he new electronic oscillators make it possible for the musician to create any wave-form, timbre or tone effect he desires, and they afford a delicacy of control and touch, undreamt of with our present gross musical mechanisms which are operated by pounding, scraping or blowing!" rhapsodized the staff writers at *Electronics* magazine about "the new synthetic music of electrons."[13] Invented by Leon Theremin in the 1920s, the instrument converts an electrical impulse to an audio signal alone.

The theremin is one of a number of musical instruments developed in the 1920s that converted the oscillations of an electronic signal into sound. In addition to the theremin, the ondes Martenot (Martenot waves) is also still in

Fig. 2.10:
VARIAN ASSOCIATES.
VISIBLE LANGUAGE: THE
UBIQUITOUS WAVY LINES THAT
SIGNIFY ELECTROMAGNETIC
WAVEFORMS ARE ADAPTED
TO A TYPOGRAPHIC DESIGN
FOR THE LOGO OF THE
COMPANY THAT INVENTED THE
KLYSTRON TUBE (*SCIENTIFIC
AMERICAN*, 1961).

use today. More complex than the theremin, an ondes Martenot uses a keyboard like a small organ and further enables the player's control of pitch and timbre through a separately controlled wire string as well as stops and pedals. Developed in France in the late 1920s by Maurice Martenot, it allows for the creation of an extremely wide range of sounds and textures. It has contributed to the chamber and symphonic musical oeuvre of the twentieth century as well as to the avant-garde.[14] Martenot was a telegraph operator during World War I and became fascinated with the sounds that resulted from signal interference in his tube-driven electronic communications system. Following the war, he became determined to "turn the raw material of electricity into music."[15]

Both the theremin and the ondes Martenot made variable signals into durable artistic media. More instruments were developed for "oscillator" in the 1920s than those that survived to be used into the twenty-first century: A 1932 article in *Electronics*—the magazine's heyday of covering electronic arts—profiles the Rangertone organ and a Meissner organ-piano in addition to the instruments made by Martenot and Theremin.[16] Today both instruments are still built, composed for, and performed . . . and have their functions mimicked by pocket device applications that anyone can use.

### CRTS, LANGUAGE, AND COMMUNICATION

In the mid-1940s, RCA Laboratories developed a device that could read aloud to the blind.[17] Each reading device contained a miniature phototube—a cathode-ray tube that cast light onto printed text. The letters on the page would interrupt the reflected light, forming a pattern on the surface of the tube that the tube could "see" and transmit to a memory device. The returned shapes triggered a magnetic recording disk to play the corresponding letters of the alphabet. This Veterans' Administration-sponsored project was electronic synthesized "speech" at its most primitive: an analog signal system with no computer power behind it, triggered by light, seen by CRTs. In the 1950s more effective, computerized optical character recognition techniques were developed. But as with many technologies now

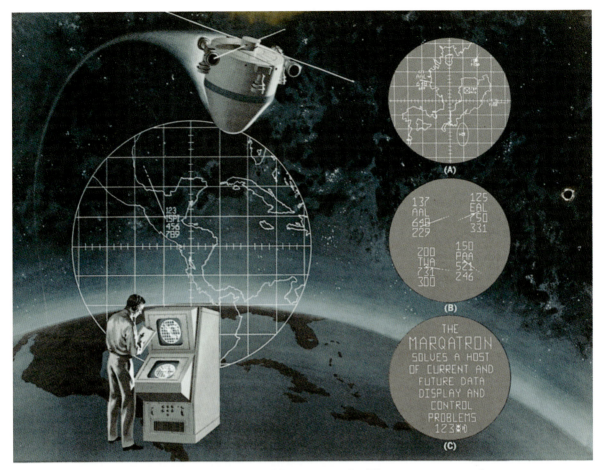

**New MARQATRON by Marquardt/Pomona produces instant visual read-out from high-speed processed data**

*Fig. 2.11:*
MARQUARDT. IN 1960, A
MONITOR INCORPORATED INTO
THE DESIGN OF A COMPUTER
WAS A NOVELTY (*AVIATION
WEEK*, 1960).

familiar to us at the pocket level, the CRT was part of its inception. In another part of the same lab, RCA researchers also developed a new CRT-based fac-simile technology using 35mm film to capture images and "photomultiplier" tubes—CRTs that amplify light—to output an image to a transmitter.[18] Devel-oped for wartime espionage applications, the cumbersome system at first had few receiving stations; years passed before it developed in the direction of the fax machine. But it did move printed pages from one zip code to another through coaxial cable, a waypoint on the road toward the electronic transmis-sion of correspondence and documents.

The conductor and composer Leopold Stokowski aimed a provocation at electronics' influence on language in 1932, suggesting a link between the new technology and written musical notation. Stokowski was an active musical futurist and spent years of his life integrating electronic instruments, including theremin and ondes Martenot, into modern orchestral compositions and performances. He had close ties to RCA, which sponsored many of his programs; he also contributed regularly to *Electronics* during its arts-friendly years. In one article he proposed replacing musical notation with oscilloscope signals. In response, the magazine's editors elevated the provocation with the comment that oscilloscopes might also be an improvement on "conventionalized ideographs" that comprise written language—the alphabet itself. "Would a child starting fresh have any more difficulty in learning . . . oscillograph wave-forms?" they asked.[19] Needless to say, no one took up the cause of replacing alphabetic writing with waveform notation. But electronic music, and the changes to all forms of music that electronic recording would bring, figure among the dominant artistic innovations of the twentieth century.

**CATHODE-RAY TUBES AND COMPUTING** In the 1940s engineers began incorporating CRTs into to the rapidly developing field of electronic computing. As a result, during the mid-twentieth century the cathode-ray tube was adapted to some uses which have since been nearly forgotten. The use of computer monitors as interfaces between people and machines took several more years to develop. Instead, the first role of CRTs in computing was in the development of computer memory.

The electrostatic tube, a specialized type of CRT, was invented in England in 1947.[20] These tubes sprayed an electrostatic charge on a nonconducting surface, usually selenium, to produce an image. Encoded data would be "programmed" into the electron discharge pattern and displayed when needed. An alternative tube-based electrostatic storage system, the Radicon tube, was produced by RCA and used a barrier grid—an alternate type of surface, not a nonconducting one.[21] Electrostatic storage tubes were used as a form of data storage for about ten years, between the late 1940s and the ascendance of magnetic core memory in the 1950s. This unusual strategy amounted to storage of information through a process of delay. The technique was similar to encoding a signal in a television tube to create a picture, but the encoded information would be stored until it needed to be retrieved. The tube would then be

Fig. 2.12:
STROMBERG-CARLSON.
THE CHARACTRON
OPTICALLY PRINTED UP TO
TEN TIMES FASTER THAN
ELECTROMECHANICAL
PRINTING (*BUSINESS WEEK*,
1957).

illuminated to display the stored data on its face. The contribution of CRTs to the history of computer memory systems is brief but noteworthy, as CRTs were again, as with art and music, the first electronics to be both component (as in a television) and device (e.g., an oscilloscope).

In this critical sense CRTs exceeded the parameters for components set by (radio) vacuum tubes. Just as CRTs both supported artistic media and were artistic media, so it was with computing. Vacuum tubes powered early computer memory but were not themselves the containers of memory, and neither were they used as an artistic medium in themselves. The long-term cultural significance of CRTs therefore amounted to more than the sum of their functions. As visual tubes and oscillators expanded the range of roles that a component could play, they extended the total reach of electronics across media types and sense domains simultaneously, foreshadowing the multi-purpose devices whose screens are their descendants.

*Fig. 2.12* is an ad from Stromberg-Carlson, the computing division of General Dynamics, one of the largest contractors of services and equipment to the U.S. armed forces. In the image, a giant tube is shown floating dramatically above the heads of the knowledge workers. It depicts a moment in history when CRTs, in this case a specially developed tube called a Charactron, were part of a computer's data output system. Electronic information would be fed to the tube from the computer's main memory bank and would be projected outward, onto the face of the tube. The illuminated tube would in turn act as a projector and the images on it would be picked up on a selenium drum and then printed onto paper or other material. A large electronic computer could print 85,000 words per minute in this way, much faster than any other printing technology of the time. In 1957 most mainframe computers were used in the data-intensive environments of industry, the military, and civilian government. The man in uniform, at left in the illustration, reinforces this sense of the target audience for this advertisement, which ran in *Business Week*. Linotype-Paul, a UK company, later adapted this technology for its Linotron typesetting machines, which were used until the late 1980s.

By the mid-1960s, CRTs had pressure-sensitive screens, allowing people to write on their surfaces with specialized light pens. Engineers used these specialized CRTs (a technology referred to as "user-visible process") connected to a computer, as a tool with which to draw and develop circuit diagrams. Two engineers could collaborate in real time using the screens, drawing, modifying, and erasing computer circuits as they discussed and tested different functions on the attached computer. The engineering literature from 1967 about "user-visible process" also uses the term "on-line" to describe the process of live, computer-mediated collaboration—perhaps the earliest use

*Fig. 2.13:*
LAFAYETTE RADIO CATALOG, 1959.

# NORAD ON THE ALERT

## Inputs from BMEWS Provide Instantaneous Missile Data Direct to NORAD Headquarters

From our vast outer defense perimeter, over thousands of miles, to the nerve center of the North American Air Defense Command at Colorado Springs, the most advanced concept of data handling and checkout is being utilized in the BMEWS system. The stakes are high, for the purpose is defense of the North American Continent.

At BMEWS installations operated by USAF Air Defense Command, computers read out missile tracking data from giant radars. This information is simultaneously relayed to NORAD's Combat Operations Center.

The Radio Corporation of America is prime systems contractor for BMEWS. At the COC, RCA's Display Information Processor computing equipment automatically evaluates missile sightings, launch sites and target areas. By means of data processing and projection equipment installed by RCA and a team of other electronics manufacturers, the findings are displayed on huge, two-story high map-screens in coded color symbols, providing the NORAD battle staff with an electronic panorama of the North American and Eurasian land masses.

The handling of BMEWS inputs at NORAD is an example of how RCA data processing capabilities are assuring the high degree of reliability so vital to continental defense.

*Out of the defense needs of today a new generation of RCA electronic data processing equipments has been born. For tomorrow's needs RCA offers one of the nation's foremost capabilities in research, design, development and production of data processing equipment for space and missile projects. For information on these and other new RCA scientific developments, write Dept. 434, Defense Electronic Products, Radio Corporation of America, Camden, N. J.*

The Most Trusted Name in Electronics
RADIO CORPORATION OF AMERICA

of this term in a context similar to the one we know today.[22] In this degree of interactivity and screen-based visuality, it is possible to see the antecedents of everything from video games to contemporary data visualization and multimedia environments.

**THE PANOPTICON**     The century-long process of development eventually obscured the common points of origin between early television tubes and other kinds of monitors. To the extent that the history of architecture can be simplified as the emergence, through time, of the window, then similarly the history of electronic tubes can be simplified as the emergence, through time, of the screen. From the perspective of the tube and its emergence, the Cold War was a great big brightly-lit opportunity for growth. Cathode-ray tubes are tubes for watching, and the Cold War was the first great war of watching, a war of surveillance, a war "fought," yet not fought, at very long distance. The new array of distance-vision technologies was crucially important to those in war work.

The cover artwork for Lafayette Radio's annual 1959 catalog is a surrealist panorama that elides the differences in scale between a spiral galaxy and a proposed space station. The space station is at the top of a diagonal stack of spherical shapes: first Earth, with "stereophonic sound" cresting the horizon, then the spiral galaxy, and then the station. Off to the right is the fourth sphere, the moon. Pinning the stack together is a pair of electromagnetic waveforms drawn in by two earthbound radar dishes. Underpinning the structure is the emergent screen array. A man operates a console that features three different CRTs: a television monitor, a radar screen, and an oscilloscope. The television is displaying a rocket prepared on a launch pad. Note that although the console is a computer, none of the screens is a computer monitor as we would think of it. The oscilloscope is displaying quantitative data of some kind, but is not yet a window onto the computer's processing systems. At this point in time the computer would have been outputting data to paper tape or printing to paper.

The potent symbolism of a new "eye" in the landscape, combined with the incipient cultural dominance of the television screen, inspired artists to use increasingly finely-rendered surrealism to suggest the new relationship between humanity and machine. Over time the "eye" got larger—much larger. The upsizing of the monitor is in sharp contrast to the downsizing process that other electronic components were subject to in the second half of the twentieth century. Computers got smaller. Vacuum tubes would give way to the transistor, and the miniaturization would accelerate. Radios got

*(opposite) Fig. 2.14: RCA (PROC IRE, 1961).*

smaller, as a result. All electronic devices became progressively smaller in the postwar era, except cathode-ray tubes. Eventually the "eye" became a wall. In *fig. 2.14*, projections from CRTs onto wall-sized radar screens at NORAD (North American Air Defense) are a Cold War specter of global surveillance. An RCA-made screen-projection system monitored signals from the Ballistic Missile Early Warning System, a radar array with massive antenna archi-

tecture in Alaska. Installed at the NORAD headquarters in Colorado in 1958 and 1959, these screens inspired the War Room in Stanley Kubrick's film *Dr. Strangelove*.

Less recognizable (relative to *Dr. Strangelove*, in any case) is the creatively stylized version of similar monitors created by the artist Steve Chan for ITT (*fig. 2.15*), the International Telephone and Telegraph Corporation. ITT and its subsidiary, the International Electric Corporation, were among the most secretive contributors to electronics development, but ITT did promote a wall display array that it created for Strategic Air Command, or SAC.

The time between the oscilloscope and the NORAD and SAC wall monitors was the century in which the "eye" became the panopticon. Late in the Cold War, the transformation began to reverse course, as cameras became smaller and more ubiquitous. Eventually, the twenty-first-century network of surveillance modes has come to embody the modern panopticon more than did the Cold War–era macro-monitors. This is the ultimate implication of the potential of kinescope technology: the realization of the panopticon, or vision of a future of total surveillance, first posited by the nineteenth-century English philosopher Jeremy Bentham.[23] Simultaneously, as the great eye of surveillance watches us, kinescopes have also enabled every screen-based form of engagement that currently holds the world's gaze.

## CHAPTER THREE

# COLD ROCK, WARM LIFE: CRYSTALS

It has been found that the frequency of vibration in [a] piece of quartz is extraordinarily constant and that it is very useful as a radio standard. In association with a small electron tube it acts as an oscillator or generator of a current. . . . As the frequency thus produced is accompanied by numerous harmonics, the crystal is a standard giving several radio-frequencies.

—"Radio Uses of Piezo-Electric Crystals," 1925[1]

**PRELUDE** In 1818, Mary Shelley wrote the first modern science fiction novel, *Frankenstein, or the Modern Prometheus*. Shelley's Dr. Frankenstein captures lightning in an electric circuit and channels it into the body of the creature he has built, bringing it to life. In her use of electricity to jump the gap between life and death, Shelley was drawing upon a deep human fascination with electricity. In uncanny forms like sparks or lightning, electricity can change shape and move through the air independent of wind or other visible external forces. It appears to "live" while it is in motion, and to "die" when it is extinguished. Captured and trapped within wires and circuits in the nineteenth century, electricity seemed an almost unnatural force. It powered technology, yet it resisted being categorized as a mechanical thing. Nineteenth-century medical revelations that the beating of the human heart is powered by an electrical charge within our bodies further closed the distance between organic systems like the human body and the lightning bolt.

Electronics is the technology of harnessing electrons—the constituent elements of an electrical current—and using them to make currents do differ-

ent kinds of work. Electronics extends the science of electricity, though the immense technological potential that is opened up by the control of electrons forms a qualitative distinction between them. The common starting point with electrical current places electronic technologies in a liminal category that is opened up by the lifelike qualities of electricity. If people did not quite know how to classify electricity, the potential applications of electronic devices could also be hard to imagine. Navigating the new technology, artists had a lot of leeway in slipping electronics into a permeable boundary layer between human and machine.

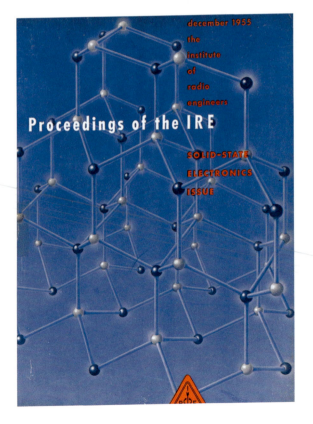

## CRYSTAL STRUCTURE

The next scale up from the atom is the arrangement of atoms into the structures that form solid matter. In the 1890s, the German scientist Max von Laue discovered that if an x-ray beam (from a cathode-ray tube) was projected through crystal, the light was dispersed in a geometric pattern. The nature of the projection also indicated that light moved in waves, a quantum revelation of light as both wave and particle. The geometric nature of the projected pattern had its own additional significance: the innate symmetry of the molecular structure, suddenly visible to the naked eye, would be the key to a new world of uses for crystals.[2]

The English scientist and autodidact Michael Faraday had already theorized in the 1830s that magnetism was a characteristic possessed by each atom within solid materials. When scientists paired Faraday's existing theories of atomic electromagnetism with new physical observations of the latticework pattern of crystal structure, a deeply new

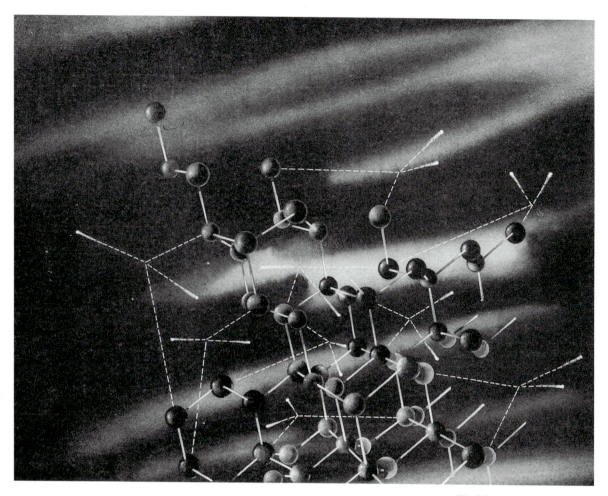

*Fig. 3.3:*
IBM. THIS AD WAS SINGLED OUT
BY *INDUSTRIAL MARKETING*:
"THE ILLUSTRATION, WHILE
FASCINATING, IS A COMPLETE
PUZZLE, BUT FROM WHAT WE
KNOW ABOUT ELECTRONIC
ENGINEERS, WE'D GUESS THIS
IS A VERY GOOD AD.... GOOD
INTELLECTUAL ADVERTISING"
[*PROC IRE*, 1959].[4]

understanding of the electronic, and magnetic, cohesive nature of matter itself began to emerge. Magnetism originates in the properties of atomic particles (electrons), and is the force that holds them together in unique atomic structures: the elements. It is one of the three energy fields that govern all matter: smallest is the nuclear force that binds protons and neutrons inside the nucleus of an atom; largest is the gravitational force that binds our feet to the ground and planets to solar systems.[3] The discovery that atoms form a symmetrical latticework within crystals shook science and inspired worldwide attention to the newly visible atomic realm.

In the early twentieth century, following the discoveries of Faraday, von Laue, and other scientists, the English scientist J. J. Thomson theorized the existence of subatomic particles, electrons. When Thomson's theory was linked

with the phenomenon of electromagnetism, the picture of subatomic matter became clearer,[5] an image clarified by the Rutherford-Bohr "solar system" model of the atomic structure. Nineteenth-century experiments with x-rays and crystals revealed that electrical energy moves in patterns and channels established by the structure of the atoms *within* solids. Electromagnetism is the energy that binds atoms together to form matter.

Crystals were the initial focal point of electromagnetic research because of their opacity and the highly symmetrical nature of their atomic structure. Scientists began studying how to manipulate these energy patterns. In this context the terms "crystal" and "crystalline" have two closely related but different meanings. "Crystalline" refers in general to the pattern of organization of the atoms within almost any solid material. Most solids possess symmetry to some extent, but in this case the term is neutral, referring to the structure itself no matter its degree of symmetry. Quartz crystal, on the other hand, is a particular stone that possesses a very high degree of symmetrical order in its molecular structure. Quartz is typically translucent and can be mined raw from natural resources or "grown" in a laboratory to specific dimensions.

Both crystal structures and quartz crystals are essential to electronics, because within solid materials the degree of symmetry of the molecular structure, and the shape and density of that structure, affect how those materials can be used within electronic devices. *Crystal* and *crystalline* also offered a portal to artists through a unique physicality; crystal structures are the networked extensions of the single atom, mimicking cell structure and other regularly designed systems. The crystal stone itself is a naturally modern subject, its aesthetic value derived from its multiple clear and symmetrical faces (*fig. 3.4*). Artistic exploration of the symmetrical nature of crystals and crystalline materials defines one entry point of angular geometric forms in the long dialogue between art and electronics.

In 1880 the brothers Pierre and Jacques Curie discovered that crystals would emit voltage when pressure was applied to them. Pressure loosens free electrons within the crystal and channels their movement through the latticework of the atomic structure and out into the air as a charged electromagnetic (EM) emission, an electronic current—a phenomenon called piezoelectricity. Quartz in particular naturally emits stable current in EM frequencies that are within radio broadcast range. The thicker the quartz crystal is sliced, the lower the frequency of electromagnetic radiation it will emit when squeezed; the thinner it is sliced, the higher the frequency.[6] Because of these properties, quartz crystal was put to use as early as the 1920s in transmitting radio signals and setting standards for radio frequencies.[7] Vacuum tubes may have amplified early radio signals, but quartz crystals controlled the frequencies at which they were broadcast.

**SONAR**     Piezoelectricity is most powerful within symmetrical atomic structures; for that reason, quartz crystal was appreciated early in the twentieth century for its usefulness in telecommunications. A system using the piezoelectric properties of a crystal can be designed to emit particular audible frequencies, or ultra-audible frequencies (those too high to hear). In 1915 the French inventor Paul Langevin designed the first device to harness this potential for a practical purpose. The device used quartz crystals to emit a beam of ultra-audible sound waves under water. Upon return, the resulting echo pattern yielded an acoustic map of solid objects. Initially designed to guard ships against icebergs, the new technology was doubly useful against submarines during World War I, and became known as sonar.[8]

In the artwork accompanying an advertisement for Stromberg-Carlson, a division of the military contract firm General Dynamics (*fig. 3.5*), a listening ear descends through water, touching the periscope "eye" of a submarine. Where the angular reaches of ear and eye meet in the water, they form a four-sided "crystal" of perception. (While true crystals have six sides, their angular, symmetrical nature is easily suggested by angular designs such as this one.)

### ORGANIC, OR MECHANICAL?

Crystals were the subject of deep laboratory exploration during the interwar period, when scientists discovered that they could be "grown" in the laboratory. By 1934, crystals were being used as phonograph needles to transmit sound (one product, produced by the Proctor electronics firm, was called the "Proctor Piezo phonograph reproducer").[9] Caught on film, the process of crystal growth is slow, and inescapably similar to the organic growth of cells.[10] At the same time, crystal structure was discovered to have "handedness," as it was characterized in early years of research. Even the most perfectly formed crystals possess microscopic degrees of asymmetry, without which they are unable to emit the desired EM charge.[11]

All usable crystals are either "left-handed" or "right-handed." Those that are truly, utterly symmetrical, or ambidextrous, lack piezoelectric properties. Human beings are prolific categorizers of the objects in their world, and one of the most general points of organizational distinction that we make is whether a thing is living or not living. Properties such as "handedness" interrupt the smooth categorization of crystals as nonliving things by introducing biophilic aspects to their nature. They seem "alive" because they "grow," and because they share some quirks with living organisms. This biophilic naturalness was a step beyond the similarly confusing lifelike nature of electricity. As "alive" as a bolt of lightning might seem, it is not necessarily life-affirming. The growth of crystals through a process that looks like cell division (really the multiplication of the atomic latticework) adds a warmth to our understanding of them. The electromagnetic "hum" of crystals also inspired attention from spiritualists, who saw the stones as a source of energy beneficial to human beings.

THIS ELECTRONIC WORLD

*Fig. 3.6:*
HERBERT BAYER FOR
GENERAL ELECTRIC (FROM
"ELECTRONICS: A NEW
SCIENCE FOR A NEW
WORLD," 1942).

The U.S. Army Signal Corps, the radio communications division of the army, contracted with Reeves Sound Laboratories in the early 1940s to mass-produce crystals for wartime radio use. This laboratory, whose work is detailed in the industrial film *Crystals Go to War* (1943),[12] relied for its supply on crystals mined in Brazil. The film demonstrates that naturally occurring crystals are typically large, stony objects with an irreproducible heavy physicality. For refined industrial use, scientists would "grow" crystals in laboratories by combining individual crystal grains with a base substance.[13]

In 1932 the electromagnetic vibration of quartz crystals was tapped to regulate the movement of a new kind of timekeeper, the first quartz clock.[14] The electromechanical device, developed at Bell Laboratories, enhanced the accuracy of timekeeping by a thousand times over the previous standard, the pendulum clock. Crystal technology remained the standard of precision for about twenty years, between the early 1930s and the mid-1950s, when the quantum movement of electrons was harnessed to timekeeping in the atomic clock. As an avenue to superhuman precision, crystals and their properties were identified with pure technological efficiency—a bit of an irony given how much they resisted being understood in mechanistic terms.

Crystals are the centerpiece of *fig. 3.6*, an interior page of Herbert Bayer's booklet for General Electric, "Electronics: A New Science for a New World." The artwork depicts a crystal garden that is both of the planet and not of the planet. This particular collage manages to express virtually all of the categorical confusion and excitement that crystals arouse. Crystals are shown in both forms: a naturally occurring cluster, the bright-white dominant image on the page, together with sketch outlines of more perfect laboratory crystal structures. The crystals are surrounded by the gentle organic forms of roses, a fern, a starfish, and a butterfly. A sky-gazing human baby anchors the entire composition. High above, a Rutherford-Bohr model of the atom floats amid the starscape, mirrored by a spiral galaxy, the softest and most "organic"-appearing of constellation types. The atomic model in the picture bridges the realms of organic and inorganic, as it represents the constituent element of all matter. Together with it, the carefully set pair of inorganic elements—crystal and nebula—express three dimensions of scale: the microscopic, the tangible, and the astrophysical, which are associated with the three energy fields of matter. They are in turn shown to be "at home" within our familiar, earth-bound world.

This artwork places crystals at the center of both a scientific frame—the scalar relationship between the electron and the solar system—and a cultural one—the changing status of electronic technologies in relation to the human experience. The most visible job of industrial communications is to educate others and explain new technologies. Less visible is this ancillary necessity of naturalizing those technologies. Underlying the drive to make cultural sense of electronics is the open question of whether electronic technologies are "natural." Are they an extension of our human engagement with the world, as natural as a rose garden, or even a baby? Or are they innately mechanical despite this artistic intervention?

None of the three energy fields that hold matter together are mechanical in nature; the exploration of crystals and electricity as "organic" phenomena continues the process of the model-making of scientific phenomena of which the planetary model of the electron was a part. The invocation of organic forms in graphic terms by artists is a small part of a much wider process that played out over the twentieth century, in which physics and other sciences gradually became understood in mechanistic terms.[15] Commercial artists addressed that question obliquely, in order to both explain and naturalize the new technology. In the aggregate, their mandate to make new technologies appear friendly to human beings caused the resulting artworks to lean toward biophilic and other natural frames of reference. Like lightning, crystals could occupy a liminal space between warm life and cold rock, and the artwork created to explain them exploits that space to its full extent. That the chief value of crystals is the ability of their electrons to sort and channel electrical current makes the comparison to lightning even more apropos. Bayer has positioned the new crystal-based technologies within a context so familiar that the artwork goes beyond the operational task of naturalizing new technology and evokes a human sense of hope for the future.

**WALTER MURCH FOR SCIENTIFIC AMERICAN**

Unlike other electromechanical components, such as switches and wires, crystals are visually familiar, even iconic. Quartz crystals are silicon dioxide, a rock element with a geometric structure familiar to a non-technical public from objects such as table salt, decorative stones, and geological formations. The elemental nature of the

*Fig. 3.7:*
WALTER MURCH FOR
*SCIENTIFIC AMERICAN* (1949).

crystal structure makes it something of a classical subject for visual art. Six years after Bayer's work for GE, the artist Walter Murch[16] (1907-67) approached the crystal as a subject for a still life. *Fig. 3.7* was created to illustrate *Scientific American*'s first feature explaining the piezoelectric effect to the public.[17] The painting, in oils, combines elements of surrealism and neoclassicism. A table is covered in green cloth, its texture and appearance in harmony with the tablecloth of the seventeenth-century Dutch masters. Hovering over the table, in defiance of gravity, is a giant crystal. Its position suggests that it is both natural and supernatural.

The explanatory copy reads, "The floating crystal is ethylene diamine tartrate (EDT)," the piezoelectric properties of which were used as early as the late 1940s in telephone systems. "At left in the painting," the magazine copy continues, "is an EDT crystal filter unit, which separates carrier frequencies so that many frequencies can be transmitted simultaneously over one telephone cable. The gold-colored element at the center of the unit is a crystal painted with real gold; at far right, a fully assembled unit lies on the table. These constituent crystals were developed at Western Electric as a synthetic substitute for natural quartz." EDT could be grown from a "seed" of quartz crystal. At Bell Laboratories the crystals were grown in specialized tanks, forming sparkling clusters that stretched to several feet in length.[18] In Murch's painting, the "seed" is represented by the cloudy right quarter of the floating crystal.

Walter Tandy Murch was a Canadian painter whose career in commercial art in the 1930s and 1940s was a prologue to a fine arts career that lasted until his death in 1967. He was at home making portraits of everyday objects; his first subjects were the material artifacts of everyday life on the Canadian prairies. The son of a clockmaker, his early interest in mechanical objects prepared him to create refined, evocative depictions of the works of mid-century science and technology. Although Murch expressly rejected a total identification with the surrealist movement, he was nevertheless strongly influenced by it, and elements of a surrealist approach turn up consistently in his work.[19] Familiar elements are often unstuck from their landscapes and float together in new compositions of association, as they do in *fig. 3.7*.

Murch had a special sensibility for the tension between organic and mechanical forms. Many of his paintings of machines include a vegetable or a piece of fruit posed on the table. In *fig. 3.7* the organic is only hinted at, in the artist's use of a soft green tablecloth to underlie the crystal subjects, a setting closer to a kitchen table than to a laboratory. In the late 1940s Murch was commissioned by *Scientific American* to create a series of cover paintings, of which this painting was the first.[20] His style made him an ideal artist to capture public interest in favor of emerging technologies. His 1949 portrait of piezoelectric crystals became the front cover of a boxed set of specially printed plates of artwork made for the magazine, released to the gift market in the 1950s as *Art and Science*.

**MICRO-PHOTOGRAPHY** Artwork about crystals takes two main forms: it is focused either on the geometric physicality of whole crystals, as in *figs. 3.6* and *3.7*, or it is based on the microphotography of crystalline structures. As long ago as the 1830s, Japanese scientists studied the crystalline structures of snowflakes under a microscope, and Japanese artists drew inspiration for textile designs from those photographs.[21] Microphotography, beginning in the 1920s, opened up a wider world of geometric forms to inspire both artist and architect. The new science of studying the internal atomic structure of matter was called crystallography. It became a very important physical science in the early twentieth century, as the range of properties and values expressed by different atomic structures was gradually realized. The electron microscope—powerful enough to view the inside of an atom—was invented in the early 1930s in Germany and was quickly adopted around the world, vastly enhancing the sensing power behind microphotography. It did not take long for the artistic implications of these new forms to be explored. In 1951 the Festival of Britain sponsored a Crystals Design Project, coordinated by the crystallographer and designer Helen Megaw. Megaw invited dozens of craft and industrial designers to create tableware and textile designs for the festival, all based on crystallographic patterns from her research laboratory.[22]

As Cyril Stanley Smith writes in the 1965 volume *Structure in Art and Science,* edited by György Kepes, "Although for centuries man has been fascinated by the geometrical shape and glitter of natural crystals, he has only recently come to see that the essence of crystallinity lies not in external shape but in the uniformity of the relationship of atoms to their neighbors within the crystal."[23]

In *fig. 3.8*, a honeycomb design is extrapolated from the crystal structure of materials being tested for strength by the Sandia Corporation (now Sandia National Laboratory). In this design, the dual geometric and organic nature of crystal structures forms the basis of an abstract painting. The flush-right block of serif text emphasizes the irregular form of the structures; later in the book, we will see language and letterforms assume an even larger role as graphic elements in themselves.

Sandia was established as a government research laboratory within the Manhattan Project. After the war, the laboratory was turned into a civilian corporation and upon request from the army it was overseen by AT&T's Bell Laboratories throughout the Cold War. Then, as today, research at Sandia was focused on microscopic study of crystalline structures, or materials science, a science with broad applications both within and beyond electronics. Sandia's New Mexico and California laboratories have developed materials for a range of industrial applications, including metal alloys for industry and war production. Like the panopticon made possible by enormous CRT projections, this Cold War–inspired application of electronics-oriented research lost no traction in the millennial era; after the Cold War, ownership of Sandia National

TWO-PHASE SHOCK TEST

At Sandia, scientists are currently engaged in producing, measuring and interpreting shock phenomena . . . effectively expanding the scope of this field through the use of new techniques and facilities. Shock phenomena is but a single area in Sandia's broad, continuing program of materials development and applied research.

Sandia presently has openings at the PhD level in the following fields: Materials Science, Engineering, Mathematics, Physics, Chemistry and Ceramics. Qualified scientists interested in careers at Sandia are invited to send resumes to Professional Employment Section 569.

SANDIA CORPORATION

ALBUQUERQUE, NEW MEXICO
LIVERMORE, CALIFORNIA

An equal opportunity employer.

Laboratory was transferred to Lockheed Martin, the largest twenty-first-century contractor of U.S. military hardware, for ongoing war work.

**FERRO-MAGNETIC DOMAINS** While Bell Laboratories was dominant from the 1930s through the 1970s, materials science research was conducted at different levels by most major industrial companies engaged in electronic development. At General Electric, the electronics laboratories conducted extensive research into the magnetic properties of crystal structures. The flow of electrical current through a substance determines its usefulness as a conductor or insulator within an electrical or electronic device. This flow is controlled by the specific magnetic properties of a substance's atomic building blocks. Ferrites are magnetic oxide materials that occur in the stone magnetite (from which the term "magnet" is derived), and also in iron, nickel, and other minerals. They are significant because they possess extremely high electronic resistance, meaning they can be put to work controlling the flow of electrons in ways useful to radio, telephone, and telecommunications engineering (at first, and later to all electronic components).

Within ferrites—iron—the crystalline structure is magnetized more than in other types of materials, and the variations in magnetism within different territories of a single slice of crystal are referred to as *magnetic domains*. Each magnetic domain affects the material's electronic resistance, and therefore bears on the material's use in electronic components and other assemblies. In 1951 György Kepes wrote about magnetic domains and their relationship to art in *Language of Vision*, noting that "color, value, texture, point, and line . . . radiate different amounts of energy"[24] in drawing a link between the practice of painting and the properties of electromagnetism. In the course called Visual Fundamentals that Kepes taught at MIT at the time, he encouraged his students to extrapolate paintings from microphotographs of magnetic field forces.

In the 1950s and 1960s General Electric had an in-house art department that produced renderings of the research laboratory's products and processes, and other aspects of the company's work, for a variety of purposes. The company contracted out most of its mass-market advertising and special projects such as the 1942 booklet, but it kept at least one staff artist, Ken Staley, busy making artwork about the research of GE scientists. In 1965 Staley was given a photomicrograph of research into the ferromagnetic domains in "thin single-crystal nickel platelets" conducted by the physicist Dr. R. W. De Blois in GE's research laboratory, and was asked to use it as the basis of an illustration (*fig. 3.9*). As in Ken Staley's work for GE, in Jacqueline

*(opposite) Fig. 3.8:* THOMAS HOLLAND FOR SANDIA CORPORATION (*SCIENTIFIC AMERICAN*, 1961).

Casey's work for MIT's Lincoln Laboratory magnetic domains appear under an electron microscope as geometric sections within a crystal surface, and more closely as fine hairlike subpatterns within crystal structures (*fig. 3.10*). It's impossible to know the extent to which either Staley or Casey was familiar with Kepes's formulation of the value to art in magnetic domains, but their works exemplify his observations.

Staley's painting first appeared on the cover of the issue of the *Journal of Applied Physics* in which Dr. De Blois's research was reported. Its shapes and lines are drawn directly from De Blois's photograph of the walls between different magnetic domains in the nickel platelet. The painting gained wider exposure when it was exhibited at the Albany (New York) Institute of History and Art's "Art in Science" show in the fall of 1965. Staley proudly remarked in correspondence that the painting was one of twenty-five items chosen for the exhibit from over 156 art objects collected for the show's catalog. The exhibit, supported in part by the Electron Microscope Society of America and juried by György Kepes among others, skewed heavily in the direction of "artistic" microphotographs of cellular and crystalline structures. But it also included a number of interpretive works by fine artists, such as Staley's enamel painting.[25] Unlike artists who made artwork for the purpose of external communications, Staley was accustomed to making work to communicate internally, within GE, and it was a big event for him to exhibit in a gallery. Then, much to Staley's surprise, his painting of ferromagnetic domains gained traction in the modern art world. It won an award within the show, and when the *New Yorker* covered the show, art critic Robert Coates singled out Staley's painting, remarking on its "solidity and strength."[26] Six years later, the painting was appropriated into a postmodern screen print collage titled *Bash* (Baroque All Style High) by the Scottish-Italian artist Eduardo Paolozzi.[27] The story of this painting is among the more unusual and detailed journeys of an artwork featuring boisterous colors—crossing some wide boundaries between the worlds of sponsored art and fine art.

I doubt it is a coincidence that both appearances of Jacqueline Casey's *Magnetic Domains* on the back cover of the *Proceedings of the IEEE* occurred in the month of November (1965 and 1966). Illustrating a recruitment advertisement for MIT's Lincoln Laboratory, her interpretation of the delicate fronds of domain-shaping irregularities in a crystal surface resemble nothing so much as crystals of frost or ice. Like Staley, Casey was a staff artist; however, she was with MIT's noted graphic design department rather than on staff with the laboratory. In her sixteen-panel composition (*fig. 3.10*), Casey has imposed a geometric regularity on these particularly irregular magnetic domain patterns.

A semiconductor is any crystalline material that allows some electrons to pass through it while impeding others. All semiconductors sort electrons, and in doing so

*Fig. 3.11:*
THIS 1945 ADVERTISEMENT
FOR A SEMICONDUCTOR
MANUFACTURER IS UNIQUE
IN ITS INVOCATION OF THE
TRADITION AMONG WOMEN OF
PASSING CRAFT KNOWLEDGE
TO ONE ANOTHER ACROSS
GENERATIONS; HERE, THE
CRAFTERS ARE ELECTRONICS
WORKERS (*PROC IRE*, 1945).

alter the behavior and the properties of the electron stream within the electrical current. Research into the properties of crystal structures was largely aimed at developing semiconductors—circuit components that could modulate, stop, or amplify electrical current. Diodes and capacitors are two common semiconductor components; diodes were the earliest component that could either impede or conduct current.

Research into crystal structures eventually led scientists to the conclusion that synthetic materials would produce more controlled and robust semiconductors than naturally occurring materials like quartz. Many industrial laboratories turned toward the development of synthetic alloys: molecules of iron or steel combined with substances such as silicon, nickel, or manganese (or a large number of other possibilities). All were formulated to offer specialized magnetic domains, each with its own particular applicability to electronics.

The discovery of lattices of atoms—microscopic networks—introduced a geometric logic to the science of electronics. It was a logic that related to organic systems such as cell structures as clearly as it did to the mechanical architecture of future technologies, yet crystals and crystalline structures surpassed pure geometry to reach into the realm of art. The latticework of electrons ushered in the motif of connective patterns within a network, anticipating the coming era of networked information systems. They touched fine art, and fine art touched them back. In 1947 the study of crystalline structures yielded a wholly new invention, one with implications far beyond the reach of crystals themselves: the transistor.

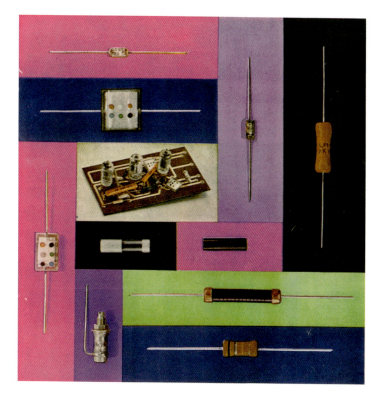

*Fig. 3.12:*

CORNING GLASS COMPANY, PROMOTING THE COMPANY'S LINE OF GLASS-CLAD SEMICONDUCTOR COMPONENTS. SHOWN ARE A VARIETY OF RESISTORS AND CAPACITORS, AS WELL AS A SMALL CIRCUIT BOARD AT CENTER. THE ELEMENTS ARE ARRANGED IN A MODERN MULTIPANEL DESIGN: ELEVEN PANELS ASYMMETRICALLY YET GEOMETRICALLY ARRANGED, MIMICKING A MODERNIST BLOCK DESIGN COMMON IN THE LATE 1950S (*BUSINESS WEEK*, 1959).

## CHAPTER FOUR

# TRANSISTORS AND CIRCUIT SYMBOLS

All in all today's variety of old and new transistors are finding their way into a staggering variety of tube and nontube
replacement equipment . . . hearing aids, portable radios,
phonographs and dictating machines, auto radios and fuel
injection systems, portable cameras, paging receivers and
instruments, machine-tool controls, clocks and watches, toys,
and even a guidance system for a chicken-feeding cart.

> — J. A. Morton and W. J. Pietenpol,
> "The Technological Impact of Transistors," 1958[1]

The transistor was the most dynamic discovery in electronics of the twentieth century. The four-year interval between its invention in 1948 and its implementation in 1952[2] marked the mid-century as a true turning point in the impact of electronics on everyday life—an impact unforeseen by the prognosticators at *Fortune* magazine less than a decade earlier. From the development of the transistor onward, electronic devices superseded the limitations of vacuum-tube technology. The transistor multiplied. It became ubiquitous. It made electronic devices small, cool, personal, and efficient. Our twenty-first-century pocket devices contain millions, even billions of transistors. In the late twentieth century the device would usher in the tide of miniaturization that continues to shape the emerging technological future.

Electronics had already begun to exert pressure on the human experience of time prior to the 1950s, as the early discoveries about the structure of the atom were associated with Einstein's discoveries about the nature of the

*Fig. 4.1:*
MELPAR ELECTRONICS
(*MISSILES AND ROCKETS,*
1959).

space-time continuum. Later the vibrations of quartz crystal brought time-keeping to a level of previously unknown precision. In the 1950s, the rapidly changing public experience of electronics brought a flood of rapid change to postwar modernity. Telephone, radio, and television all influenced the shape of everyday life, and after the war their impact accelerated. The pace with which technology began suddenly to integrate with everyday life, as a result of the transistor, met and matched those other trends. The impact of the transistor on both the public and on industry was so disproportionate to the physical impression of the device itself that depicting it demanded a novel and unique graphic strategy.

**TRANSISTOR TECHNOLOGY** In December of 1947 three scientists at Bell Laboratories discovered how to amplify—not simply transmit—electronic signals within solid materials. This was the discovery that yielded the transistor.

The transistor is a type of semiconductor. Whereas the interior of a vacuum tube sorts and amplifies an electronic signal by passing it through wafers and vacuum space, the transistor draws on the semiconducting properties of crystalline structures to produce the same effect within a solid substance. The first transistor relied on the properties of germanium crystals, which turned out to cause a type of selective sorting that created a dramatic amplification of the signal. The gadget devised in 1948 by Bell Laboratories researchers John Bardeen and Walter Brattain, under guidance from senior researcher William Shockley, consisted of two electrodes, one an emitter and one a collector, that were paired together at the top of a stack, at the center of which was a small block of germanium, and beneath it a low-resistance base electrode. The result produced signal amplification that superseded vacuum-tube amplification.[3]

"During the war, a large amount of research on the properties of germanium and silicon [crystals] was carried out by a number of univer-

**How a fragment of metal will let you dial your own long distance calls**

sity, government, and industrial laboratories in connection with the development of point-contact rectifiers for radar."[4] So wrote Bardeen and Brattain in the introduction to their first published paper about their invention. All the

research about crystallography, from the earliest x-ray experiments onward to 1948, contributed to the basis of knowledge that facilitated the invention of the transistor. The timing of the invention was therefore no accident; instead it expresses the intimate link between wartime research and the postwar electronic world.

When Bardeen and Brattain discovered the amplification property within germanium crystals, that was the discovery. When they built a rudimentary device to harness that power, that was the invention. In the case of the transistor, scientific discovery and technological invention were very closely knit, a dual and essentially simultaneous result of pure laboratory research and experimentation. The discovery was made while the two scientists were experimenting with some parameters suggested to them

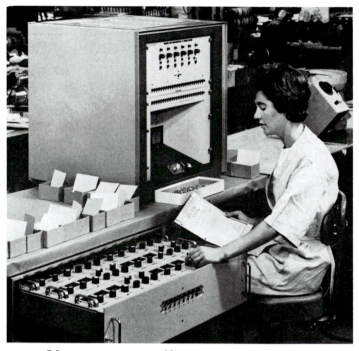

## Now... a really new s-c tester
### featuring digital programming and low cost

*Fig. 4.3:*
A TEXAS INSTRUMENTS
WORKER DEMONSTRATES
EQUIPMENT TO TEST
TRANSISTORS AND DIODES
(*ELECTRONICS*, 1964).

by Shockley, who later took equal credit for inventing the transistor. In 1956 the three shared the Nobel Prize in physics for their research.

The transistor was the single most technologically and culturally significant outcome of early experimentation with crystal structures. Its technological significance is evident; at the cultural level, it transformed our relationship to technology by shrinking it. Although it was not immediately integrated into the design of new computers, its long-term impact would facilitate the expansion of electronic technologies beyond individual devices to networks and systems.

**SHAPING THE TRANSISTOR** Early transistors were bulky by today's standards of microminiaturization. Yet they were small for their time—much smaller than vacuum tubes. From our twenty-first-century vantage point it may not be clear why the two technologies should be compared, as their uses have diverged so widely in the intervening sixty years. In the 1950s, however, transistors were a dramatic new invention that caught the makers of electronic devices off guard. Prior to the discovery, it had been thought impossible for a semiconductor to amplify electronic signals. Other semiconductor devices such as capacitors, diodes, and resistors were all already hard at work in electronic circuits, using their semiconducting properties to exert different types of control over the flow of current through a circuit. They had been studied for years, and while research continuously improved these devices, the accomplishment represented by the invention of the transistor brought this research to a higher level.

Transistors posed a tantalizing promise, but laboratory development, manufacturing, and marketability of the new device left their immediate impact unknown for a few years. Some thought the challenges of building and standardizing transistor technology would keep the device out of the market for many years. Following the 1948 announcement, Bardeen, Brattain, Shockley and their colleagues at Bell Laboratories embarked on the project of developing and manufacturing it. The laboratory began supplying the Bell System's telephone network with the first transistorized operator switches in 1952, and the next application was in hearing aids. The concepts underlying transistor technology began to be used by other companies already in the semiconductor business shortly thereafter, many of whom began to work on transistor devices of their own.

As it happened, it took only four years for industrial applications to be developed, and six years for the transistor to reach the general public. The new invention could transform and control the flow of electrons more powerfully than many vacuum tubes, and in much less space and in a much more stable manner. Transistors did not throw off heat, or overheat, in the way that had always been a limiting factor for vacuum tubes. Vacuum tube radios never really got much smaller than a toaster, and their "portability" was limited to carrying them with a shoulder harness. The transistor, by contrast, shrank the portable radio to the size of a human hand. When the first pocket transistor radio was offered to the public at Christmas 1954, it brought people into more intimate contact with electronics than they had ever been before. In 1955 RCA offered the first transistorized car radios,[5] and the term "transistorized" came to refer to jump-start modernization. Electronic devices were at once dramatically more compact, efficient, and stable than they had ever been before.

**MOVING WEST**	The pressures of the process of invention exerted a force of their own on the scientists at Bell Laboratories. By the mid-1950s the original transistor laboratory team had dispersed, and the momentum of invention shifted toward other electronics firms. William Shockley moved to California and established Shockley Semiconductor, adding to the development of the place that would become Silicon Valley. John Bardeen became a professor at the University of Illinois. Of the original transistor team, only Walter Brattain remained at Bell Laboratories. Gordon Teal, a young research scientist who was also part of Shockley's laboratory in the 1940s, left Bell Laboratories for Texas Instruments in 1952. There he invented the silicon transistor, a development that would extend the impact of the transistor dramatically. Silicon is vastly more plentiful than germanium, yet it turns out to be a similarly effective semiconductor. In addition to being more plentiful, its crystals are also easier to synthesize in the laboratory, and it is therefore much cheaper to manufacture and utilize. Teal's silicon transistors shortly became the world standard.

**REPRESENTING THE TRANSISTOR**	In *fig. 4.4*, a Texas Instruments advertisement promotes Teal's silicon transistor. The slight form of the transistor crosses and supersedes the more recognizable vacuum tube. The desert landscape expresses a few things: first, that Texas Instruments was a new heavyweight in the field following the invention of the silicon transistor; second, that the momentum of twentieth-century technology was pushing westward. These impressions were created with smoke and mirrors within the image: the saguaro cactus is not actually native to Texas; the use of the cactus to convey a sense of "Texas" was impressionistic at best, erroneous at worst. The sand stands for silicon, and the desert landscape stands in for the very "westness" of what would become known as Silicon Valley: Fairchild Electronics, the Stanford Research Institute, Hewlett-Packard, and many more. (The saguaro cactus is no more native to California than it is to Texas.)[6] Lastly, the square frame around the desert references the compositional style of the geometric modernist painter Piet Mondrian, and in doing so asserts an association between the desert landscape and the modern age.

During these early years, as with all the preceding developments in electronic technology, artists were called upon to visually interpret the transistor for both business audiences and the public. Unlike laboratory-bound discoveries such as ferromagnetic domains, the invention of the transistor

keep an eye on **T/I**

TEXAS INSTRUMENTS INCORPORATED

sand . . . heat . . . and *silicon* transistors

needed to be explained to the public—the scope of its impact on everyday life demanded it. Early depictions of the transistor were rooted to the tiny device's physical appearance. The first transistors were designed as little knobs with spindly electrodes sticking out from them. A 1950s paint-swatch modernism promoted Clevite's transistors in 1961 (*fig. 4.5*).

**DESIGN STEPS IN: THE SHIFT TO SYMBOLIC REPRESENTATION**  Early transistors resembled mechanical insects, quite different from the illuminated little room that the interior of the vacuum tube appeared to be (suggestive of the imagination), or the geometric modernism of crystal structures. Artists were stymied for a while about how to characterize them in artwork, as straight representations tended to do little to express what a technological leap the new component represented.

The device's microscopic workings defied easy graphic characterization, and its mass applicability defied its puny physicality.

While earlier electronics components such as vacuum tubes and CRTs could be easily associated with extensions of particular human senses, the transistor eluded such associations. The reach of transistorized electronics was too

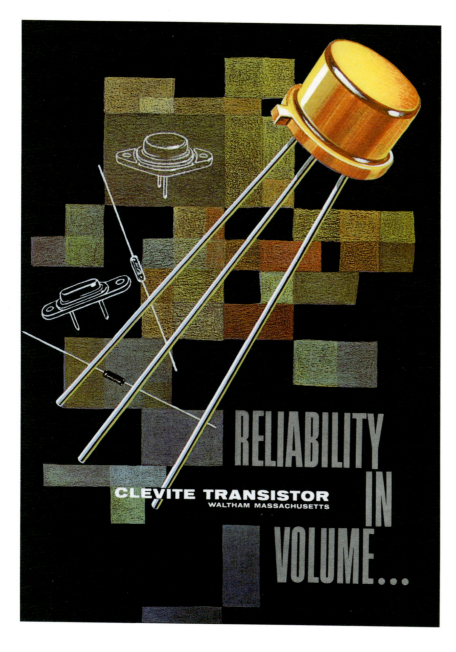

*Fig. 4.5:*
BETWEEN 1960 AND 1966, CLEVITE TRANSISTOR OWNED WILLIAM SHOCKLEY'S POST-BELL LABORATORIES FIRM, SHOCKLEY SEMICONDUCTOR. THE HATPIN FIGURES IN THE BACKGROUND ARE DIODES, ANOTHER SEMICONDUCTING DEVICE USED IN ELECTRONIC CIRCUITS. THIS AD WAS PROBABLY PUBLISHED TO REINFORCE THE COMPANY'S THEN NEW ASSOCIATION WITH SHOCKLEY (*ELECTRONICS*, 1960).

**BROADEN DESIGN HORIZONS**
*with new* **pnp drift transistors**

LITTON INDUSTRIES
Beverly Hills, California

totalized, extending beyond the sum of radio, radar, sonar, and TV. By the late 1950s transistorized electronics permeated communications, transportation, infrastructure, manufacturing, and business. As the component became smaller, the reach of the technological network grew larger. The little object itself was transcended. The significance of the transistor could not adequately be represented with a stylized hand, eye, or head, because it was all of these things.

Commercial artists working for electronics firms could have had to invent a new abstract visual language with which to portray the mighty, but tiny, transistor. Had they needed to do so, there would have been a wide repertoire of strategies to draw upon. As it turns out, they were spared that job by engineers. Engineers had done their own work to develop a representation system for electronic components, the schematic symbols for circuit diagrams. Artists of the late 1950s turned toward these schematic symbols and found there a ready-made minimalist, even geometric, graphic communications system poised for their reuse. This system was almost entirely abstracted from physical reference points, its development aimed at simplifying circuit diagrams to the greatest extent possible.

Schematic symbols hew to geometric simplicity for ease of readability. Their elements are straight lines and perfect circles, and geometric variations such as triangles and loops. The classic symbol signifying "transistor" was developed at Bell Laboratories, and was based on the device's internal structure.[7] The graphic is an extremely simple "map" of the device's architecture: a circle defines the element, and the lines within it depict the element's constituent parts: the collector, the emitter, and the base (elements of an early basic layout of the device).[8] As shown in *fig. 4.8*, the collector and emitter meet at the base, the dense vertical line that joins the apex, where the two incoming lines meet. One bears an arrow, adding asymmetry.

The Cold War, which was a war of postures and symbols as much as one of weapons and policy, forms an inescapable context for artwork based

Fig. 4.8:
THE FRONT AND BACK COVERS OF A GE TRANSISTOR HANDBOOK, FEATURING ARTWORK THAT STRONGLY ASSERTS THE ICONIC STATUS OF THE TRANSISTOR SYMBOL (1960).

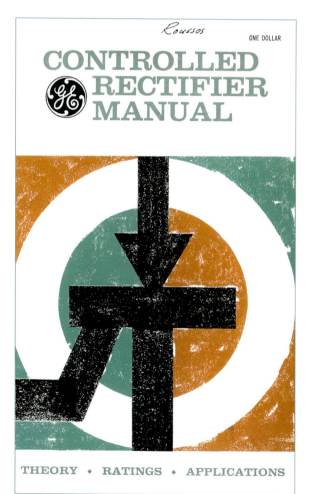

**BEHIND THIS CURTAIN HIDE THE ONLY ICBMs NOT WATCHED BY KIN TEL CLOSED CIRCUIT TV SYSTEMS**

*For full information on Kin Tel closed circuit TV, write direct.*
*Representatives in all major cities.*
5725 Kearny Villa Road · San Diego 11, California · Phone BRowning 7-6700

on the transistor symbol. The USSR anchored its wartime public image with the iconic graphic of the hammer and sickle (*fig. 4.10*). Within the smaller, regional environment of the electronics industry, the artists behind *fig. 4.9* and images like it appear to be sketching their designs with tongue firmly in cheek, offering their own war of images as a semi-satirical reflection of the wider geopolitical moment, exploiting the glancing similarity between the transistor symbol and the USSR's more widely seen national symbol.

Schematic diagrams originate with early engineers' basic need to draw diagrams of what they were doing, so that experiments and procedures could be replicated. The craft of drafting engineering plans emerged from the tradition of architectural and technical drawing. Even earlier, higher mathematics had cultivated a symbolic language that could convey spatial and quantitative concepts using numbers together with forms such as the Greek letter *delta* (△) standing for *change*.

The practice of developing symbols to stand for electrical components started with Faraday's first drawings of electrical schemes in the early nineteenth century. As the vacuum tube, the first electronic component, was developed out of the basic concept of the lightbulb, a new symbol was made for it, turning an electrical diagram into an electronic diagram. In the early 1950s the Institute of Radio Engineers (IRE) convened a joint committee with the American Standards Association, and this group of engineers developed the graphic symbol for the transistor. The work was conducted during the mid-1950s, and the symbol was first published in 1958.[9]

In the West, abstraction was introduced into art early in the twentieth century. By the late 1950s, modernist artists had developed a number of approaches to abstract representation, many drawing upon traditions that were vastly older than Western contemporary art: the use of minimalist geometric forms such as circles, squares, straight lines, and right angles. The turn toward geometric form in graphic design was striking when adapted to electronics, but its use was historically well established in the artwork of Native American tribes, as well as in the traditional arts of China and Japan and of other peoples around the world. The appearance of similar forms in durable crafts such as ceramics and weaving predates recorded history.

As a deliberate turn away from nineteenth-century Romanticism and early twentieth-century art nouveau, the focus on geometric forms dates in the Western tradition to the revolt by artists of the Viennese Secession school of design against their elders' commitment to illustration.[10] In England, the nineteenth-century Arts and Crafts illustrator Walter Crane was influenced by Japanese woodblock design and promoted the centrality of geometric form to design in his 1900 book *Line and Form*.[11] These ideas would be followed in successive phases of design evolution throughout the first half of the twentieth century. The Italian Futurist movement, Cubism, and the enormously influential Bauhaus School are all subsequent contributors to the ironclad links between geometric forms and the essence of twentieth-century modern

(opposite) Fig. 4.10:
MELPAR ELECTRONICS,
THE DEVELOPER AND
MANUFACTURER OF THE
MICROWAVE ANTENNAS
CARRIED ON THE PROJECT
MERCURY CAPSULE [*MISSILES
AND ROCKETS*, 1959].

design. Perhaps prompted by training in modern design, artists looked for inspiration to engineering diagrams, developed by and for engineers.

There they found the "new" graphic language that could communicate the ideas behind transistors. In keeping with the tradition of the twentieth-century turn toward abstraction, this language was immediately recognized and utilized. The artistic reuse, by artists, of creative work by engineers constitutes a privileged event in the mediated "collaboration" between artist and engineer that yielded the corpus of mid-century commercial graphic art. In this instance engineers—without thought to art, necessarily—created for themselves a functional but abstracted design language to communicate about electronics.

Most striking, beside the impact of the artwork itself, is the readiness with which electronics was adaptable to trends in graphic modernism. Artists did not have to reinvent the circle for the sake of electronics; instead, they merely adapted it, modifying this irreducible element of modern design to serve the unique purpose of creating a public identity for the transistor. It is not a coincidence that both modern artists and electrical engineers arrived at similar results through divergent paths. Both were responding to the presence of technology in everyday life. Early nineteenth-century diagrammers of electrical circuits were developing strategies to make their drawings reproducible—a significant difference from the fine art of the day, which was valued on the basis of uniqueness. A direct alignment between graphic art and technologies of reproduction did not arise until the advent of widespread commercial printing later in the nineteenth century. Even so, the development of geometric minimalism waited until the twentieth century, when artists deployed it in retort to the incursion of technology into everyday life, among other prompts. Then came the collision between electronics and art, and artists only had to adapt these symbols. When they did, they bridged the gap in representation between the transistor and the scale of its impact.

*Fig. 4.10* is a 1959 recruitment advertisement for Melpar, a Cold War military-contract electronics firm. Its design is a thoughtful composition by an unknown artist. Twelve of the circuit symbols most commonly used at the time are arranged in the position of the figures on a clock. The symbol for a cathode-ray tube is at the eleven o'clock position, while the transistor symbol is at six o'clock. In between them are the common graphic symbols for capacitance, resistance, ground, and the other major features of a mid-century electronics circuit.

We cannot know how deeply the artist contemplated the implications of his or her composition, but it is nevertheless significant that this artwork com-

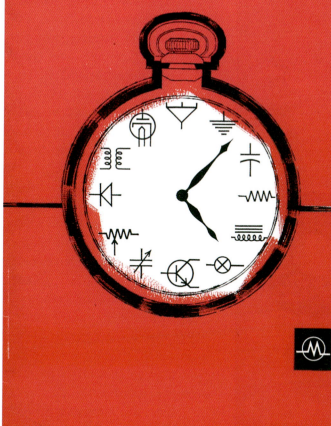

# Electronic Creativity
## is a timeless quest

✓ *Original Conception*

✓ *Design*

✓ *Expedited Production*

Melpar's deep reservoir of technological electrocraft is solving problems of almost incredible complexity—with the utmost scientific precision.

Electronic elements of the nation's space exploration and defense fall within our scope. This advanced equipment is created and produced for the U.S. government and its major prime contractors.

**MELPAR** INC

A SUBSIDIARY OF WESTINGHOUSE AIR BRAKE COMPANY

For details on provocative job openings in advanced scientific-engineering areas, write to: Professional Employment Supervisor

**3607 Arlington Boulevard, Falls Church, Virginia**

bines time with a technological story. To some extent, different eras within the century can be differentiated by virtue of the type of electronics at work and in what capacities. The emergence, in 1957, of circuit symbols as graphic design elements marks a significant shift in the relationship between technology and its representation through design: it was the moment when an abstract symbol became a more potent conveyor of meaning about a new technology than images of the thing itself.

**TUBES AND
TRANSISTORS**

Vacuum tube manufacturers naturally resisted the idea that their product was headed toward obsolescence. Tubes had been a transformative technology in their own right, had stood the test of sixty years already, and had undergone many stages of refinement, improvement, and differentiation. Long-

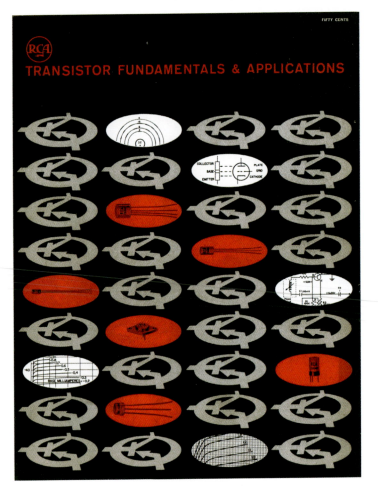

standing complaints about vacuum tubes, centered on their bulkiness and their fragility, were being addressed directly by tube manufacturers of the 1950s with new products made from metal instead of glass, and greatly compacted in their dimensions. It is in that context that *fig. 4.12*, the booklet *Tubes and Transisters: A Comparative Study*, was published: two graphic symbols standing, together, for one complex tale.

The symbol for the vacuum tube, like that of the transistor, starts with a circle, within which is an array of possible dashes, dots, and lines, each combination of which signifies different kinds of internal diagrammatic elements within the tube. Early volumes of *Electronics* magazine used the related cathode-ray tube symbol as a page design element, epigram, and general symbol standing for the entire electronics industry.

*Tubes and Transistors* is written as a value-neutral summary and guide to the state of the art in both technologies as of its 1960 publication date. It is written as if the two technologies—achieving, as they do, the same objective by different means—are as flatly comparable as an apple and an orange. But it was written and published by General Electric, a tube manufacturer from way back. From this perspective, the booklet reads like an extremely calculated expression of hope that the tube will somehow hold ground against the transistor in the 1960s. It's very convincing if you don't know the way history unfolded. You don't need to read the book to glean its angle, as the cover design deploys the two graphic symbols to bold effect. At top, the vacuum tube symbol is vivid in red, a color of dominance, while the transistor symbol perches in a secondary position below and is blue, a color associated with receptivity and passivity.

*Fig. 4.12:*
GENERAL ELECTRIC, C. 1960.

It took a little bit longer to design a computer around transistorization—transistors could not be simply substituted for vacuum tubes, as their morphology and function was too different. In order for computers to become transistorized, engineers had to start simple, with wholly new designs. In 1953 Bell Laboratories, which led the way throughout the 1950s with transistors, achieved their first working transistorized computer design. Called TRADIC, this new, small, fast computer was supplied to the air force for missile guidance and other jobs.[12] But this was only one early, isolated experiment. As the next chapter explains, transistor technology developed toward vast networks of circuits—integrated circuits—and these formed the basis of modern computers. They also offered yet another new visual sensibility, one that better conveys the relationship between transistor and computer.

**SOLID STATE** The study of the behavior of electrons within solid materials became known, after World War II, as solid state physics. Following the subsequent impact of the transistor on electronics, transistor-based electronic devices were referred to as solid state devices, to differentiate them from tube-driven devices. By the late 1960s, the phrase

Fig. 4.13:
TUNG-SOL ELECTRIC INC.
A TRANSISTOR CIRCUIT
DIAGRAM IN DELICATE OILS
(*ELECTRONICS*, 1961).

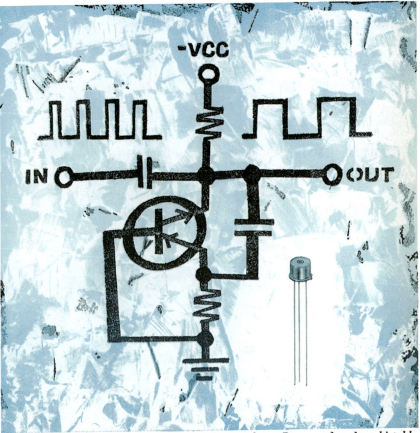

*Two resistors, two capacitors and one Dynaquad make a bistable.*

Fig. 4.14:
LOCKHEED. A TRANSISTOR
SYMBOL FRAMES A STARSCAPE
IN THIS RECRUITMENT
ADVERTISEMENT, WITH A
DE FOREST AUDION TUBE
NOSTALGICALLY PLACED
IN FRONT OF IT. AT THE
TIME, THE COMPANY WAS
CLOSELY INVOLVED IN THE
DEVELOPMENT OF BOTH
MISSILES FOR THE MILITARY
AND ROCKETS FOR THE CIVIL
SPACE PROGRAM. THIS AD
CALLED FOR ELECTRONIC
ENGINEERS TO HELP THE
COMPANY "EXPAND THE
FRONTIERS OF SPACE
TECHNOLOGY" (*AVIATION
WEEK*, 1959).

had become a popular catchphrase, and then a slogan, to indicate the latest, tricksiest, most modern tubeless devices. Through the 1970s, "Solid State!" could be seen painted on electronics store windows, advertising the technology behind the latest items. Vacuum tubes were on their way out: they were irreparably associated with bulkiness, heaviness, and fragility, while the transistor saved radios, and later everything else, from all three problems at once.

Language transfer from physics laboratories and arcane publications to popular slang is a bit unusual, and in this case it was dramatic. The phrase "solid state" took about fifteen years to move from laboratory to store window; it came from behind and overtook the more pedestrian term "transistorized." Meanwhile, the tenure of circuit symbols as design icons for the new electronic age was limited to the scant decade of 1957–1966. In *fig. 4.15*, on the cover of *PhotoFact Reporter*, an artist has turned the language into art itself, showing "solid state" as the boxy (crystal structure) wave of the future. It is held up by the same kind of disembodied but empowered hand that lifted up the cathode-ray tube in *fig. 2.7*. The appearance of the phrase "solid state" in dimensional block type invokes the tradition of concrete poetry, which uses language and typographic elements for their graphic and artistic strengths; more on that in chapter 7. This special issue of the magazine, from June 1964, was a survey of how the transistor was transforming both personal and industrial electronics.

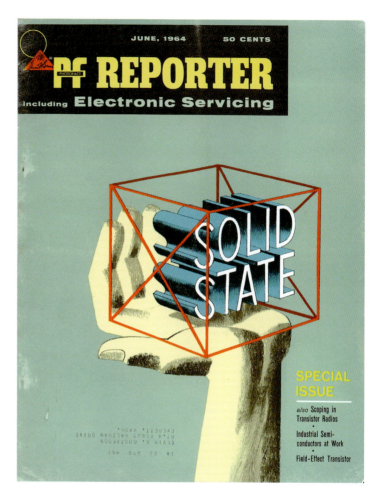

*Fig. 4.15:*
VISIBLE LANGUAGE: A LITTLE BLOCK OF CONCRETE POETRY (*PHOTOFACT REPORTER*, 1964).

In the September 1948 issue of *Electronics*, in which the transistor was announced, the magazine's editor, Donald G. Fink, made a brash proposal. In honor of the emerging era of the semiconductor device, he proposed that the family of devices be given its own nickname. He referenced the tradition of Greek-style names for vacuum tubes (audion, orthicon, pliotron, etc.), which had been started by GE technicians in Schenectady, and proposed that the new solid state devices be called "semicons."[13] For some reason this suggestion did not catch on, and the larger mouthful "semiconductor device" retained its archaic-sounding grip on the language of the next decade and a half. But then, as fast as it happened, the era of the lone transistor was over. In its place were the vast networks of transistors, the integrated circuits, that would shape the 1960s and computer-driven future beyond.

## CHAPTER FIVE

# CIRCUIT BOARDS
# AND THE MATRIX

The next evening he stayed till later, assembling a primitive circuit that made square waves out of sine waves. To his surprise and pleasure, the gadget worked the first time he turned on its power supply. He spent a happy hour varying voltages and load resistors, trying to improve an unimprovable square wave, and was obliged to concede at the end of the hour that he had hit it right the first time. This little clipper circuit had importance beyond its physical self——beyond its two triodes and their court of resistors and capacitors; it was the first electronic article Paul had ever designed, and it was the first circuit he had assembled that performed properly from the start.
—Joseph Whitehill, *The Angers of Spring* (1959),
a dramatic novel set in an electronic
engineering shop[1]

Circuit boards have a flat rectangular surface at their core, a shape that coincides with that of a painter's canvas. They connect and organize groups of components that transform electrical current and unleash the special capacities of electron streams. Active components such as vacuum tubes and transistors are surrounded by supporting components, such as capacitors and resistors, arranged to modulate and structure the current into a system—a complete circuit—that controls a radio or other device. These basic facts about circuit board structure have been true since the earliest tube-powered devices and are technically still true in spite of the intervening shifts to integrated circuits and microminiaturization. Early radio boards are the direct antecedents of today's computer motherboards.

*Fig. 5.1:*
RAUL MINA MORA FOR THE
BUDD COMPANY. BUDD WAS
THE PARENT COMPANY OF
CONTINENTAL DIAMOND FIBRE,
A MAKER OF LAMINATES FOR
ELECTRONICS (*BUSINESS
WEEK*, 1958).

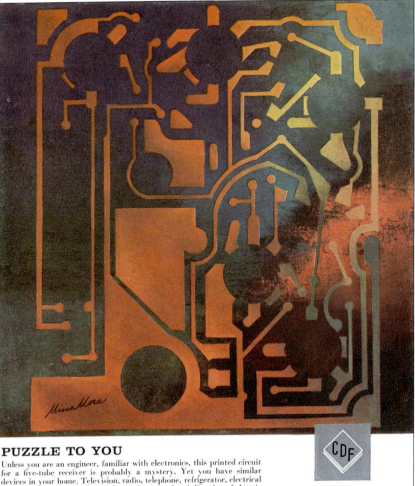

## PUZZLE TO YOU

Unless you are an engineer, familiar with electronics, this printed circuit for a five-tube receiver is probably a mystery. Yet you have similar devices in your home. Television, radio, telephone, refrigerator, electrical appliances—all have intricate circuits for control and direction of electric impulses. Continental-Diamond Fibre, a subsidiary of The Budd Company, produces vulcanized fibre, laminates and other plastics which are used for these and many other purposes by more than 3000 customers.

Budd research and development goes deep into many fields—new metals for super-sonic planes and space vehicles—nuclear systems—machines for new concepts in scientific measurement and testing—as well as our basic manufacturing for automotive and railway industries. The Budd Company, Philadelphia 32.

### MISSILES, SPACE AND CDF

Lightweight, CDF "Di-Clad®" materials for printed circuits are being used in control systems, wiring harnesses and electronic computers for missiles such as the Redstone, Jupiter, Falcon, Hawk and Sparrow. Other CDF products are used as rocket liners, flame barriers, heat deflectors and solid fuel propellant cases.

Hand-assembled circuit boards were distinctly three-dimensional. They were a high-voltage assembly of lumpy tubes, beaded components, and snaky wires, and were mounted to an interior cabinet wall of the device. As early as the 1930s, engineers realized that low-voltage radios could be produced much more quickly if any aspect of their assembly were automated. In the mid-1930s an English radio engineer named John Sargrove addressed that challenge.

Sargrove ran his own small radio factory, where he developed a manufacturing system utilizing skilled workers, a semi-automated assembly line, and a series of electronic control units that mastered the automatic aspects of assembly. The electronic control units—described in *Popular Science* as "electronic 'brains'"—used logic circuits similar to those found in early electromechanical computers.[2] When his system was publicly announced in January 1947, Sargrove became the first to propose that circuits should be treated "not as an assembly of component elements but as a 'compound' whole."[3] Sargrove saw the circuit board as a new electronic component in and of itself. It was this new component that commercial artists would soon approach as a canvas for design.

Engineers with the Army Signal Corps, the National Bureau of Standards, and RCA were working at the same time as Sargrove on their own strategies for prefabricating electronic circuits. The war had introduced an urgent need for automation and standardization in the development of radio circuits, as the Signal Corps was responsible for coordinating all military communications including "telecommunications" (such as they existed at the time, relying heavily on radio networks). By 1947 the National Bureau of Standards had a demonstration version of a printed circuit, and by 1951 the Signal Corps Laboratory had refined four major techniques for the automatic fabrication of circuit boards: printing, spraying, stamping, and etching.[4] The 1950s then became a closely framed period in the history of board development. Whereas in the first half of the century boards were handmade rather than fabricated, by 1960 they would be changed again by the development of microelectronics. In between those benchmarks were several stages of emergence for the major techniques of automation. The delicate 1958 oil painting of a printed circuit board (*fig. 5.1*) by Raul Mina Mora, who was also a children's book illustrator, captures the excitement of the circuit board decade.

Each technique resulted in circuits that were embedded in channels in a plastic substrate, replacing the hand wiring and soldering of individual

**Chart new paths in printed circuits with Panelyte®**

*Fig. 5.2:*
ST. REGIS CIRCUITS
(*ELECTRONICS*, 1960).

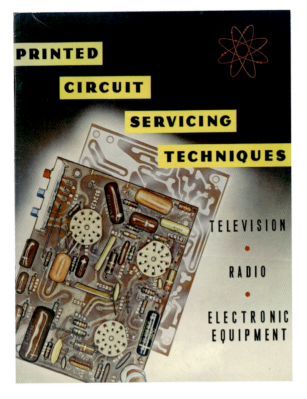

Fig. 5.3:
THE COVER OF THIS 1960 RCA
BOOKLET SHOWS DIODES
AND CAPACITORS WIRED
TO A STATE-OF-THE-ART
PRINTED CIRCUIT BOARD. THE
CIRCULAR DISKS ARE VACUUM-
TUBE SOCKETS.

components. The prefabricated circuits could be made from any of several metal alloys, with those based on copper the most commonly used;[5] the conductive metal would be sprayed or painted onto the substrate, on which a desired circuit pattern had been etched. The conductive metal would be repelled or etched away from the part of the board where it was not needed and what remained was the circuit design. The prefabricated circuit channels at first replaced only the connective wires of circuit arrays. Though the copper "wiring" was now a circuit rather than a wire, the end result was still quite three-dimensional: the other components, including vacuum tubes, would be wired to the surface of the board after its manufacture (*fig. 5.3*).

The range of production techniques for printed circuits that emerged between 1948 and 1951 brought the electronics industry into close collaboration with the plastics industry. Previously the plastic elements of electronic devices were distinct from the electronic "brain" inside of it—as distinct from each other as form (plastic) and function (electronics). With printed circuits, plastics, metal alloys, and circuit design became closely interdependent.[6]

## CIRCUIT BOARDS AND HANDCRAFT

Craft bench. Easel. Peg board. Artist's palette. The starting points of both electronic tinkering and art-making are alike. A home electronics workshop is equipped with pliers, cutters, coils of raw materials such as wire, and bins of components. It is not unlike a home craft bench, which is similarly supplied with pliers, cutters, coils of raw materials such as yarn, and bins of components such as buttons and thread.

Circuit board development was driven by the twin imperatives for automation and standardization in circuit assembly, yet the production techniques that resulted from those imperatives reclaimed some traditions of handcraft that originated in the workshops where hand-assembled circuits were developed. The philosopher of technology Lewis Mumford pointed out that until the mid-nineteenth century, "handicraft itself was the mediating factor between pure art and pure technics, between things of meaning that had no other use and things of use that had no other meaning."[7] In Mumford's sense, the craft bench was the bridge between art and applied skill, broadly conceived ("technics"). That historical context makes it particularly noteworthy that techniques first developed for the production of graphic art, such as masking and screening, were incorporated into circuit manufacturing even as automation took over the assembly of parts. Silk-screening was used to formulate resistors within printed circuits. Metal alloys were formulated to match the resistance value and voltage stability, among other qualities, of conventional resistors; in liquid form, they could be silk-screened onto the base in several steps. Before the copper circuitry was added to the etched channels, the silk-screened silver resistors would be baked, cured, and dried.[8]

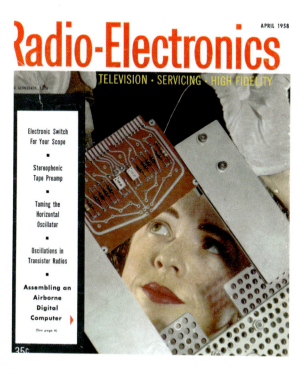

Fig. 5.4:

CATHERINE BOCHAT, AN ELECTRONICS TECHNICIAN AT HUGHES AIRCRAFT IN EL SEGUNDO, CA, SOLDERS CONNECTIONS ON THE PLUG-IN ASSEMBLY OF THE DIGITAIR, THE FIRST DIGITAL COMPUTER USED FOR FLIGHT ENGINEERING. THE DIGITAIR USED ETCHED CIRCUITRY FOR 75 PERCENT OF ITS WIRING (RADIO-ELECTRONICS, 1958).

A new type of manual work was also needed at a different stage in the production cycle: the drafting of circuit plans (called "printed circuit master drawings") which were the blueprints for new devices. These circuit plan drafters adapted their drawing boards from the making of maps and engineering diagrams to this new type of work. Circuit plans were hand-drawn, then reduced for printing before being mechanically stamped into the plastic base board. Later, photography and photolithography were used in different aspects of the manufacture of fabricated circuits.[9] These processes could be done at home, as photographers had been doing for years, and dedicated radio buffs built their circuit boards from scratch. In October 1959 a new magazine was launched to serve the fast-growing professional specialty, *Graphic Science: The Magazine for Draftsmen*. The lead article in the first issue was about graphic art for electronic circuit production, while the word "science" in the title tacitly signaled the convergence of graphic *art* with its leading client group of the day, engineering firms.

The turn toward craft techniques in association with electronics manufacturing was a boon for the Ulano graphic arts supply company, makers of Rubylith masking film. Some of their ads in *Electronics* magazine from the 1950s and 1960s feature graphic art borrowed from a craft context, showing people using their masking films to put virtual wine in virtual bottles, but then sometimes a circuit board in bright Rubylith red appears too. In the originally published edition of *fig. 5.5*, from 1966, the red masking film is offered

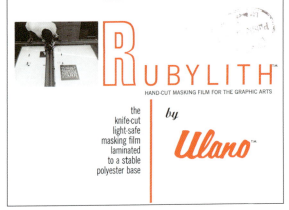
Fig. 5.5:
ULANO. ON THE ORIGINAL
PAGE, THE SAMPLE MASK FOR
A CIRCUIT BOARD (IN RED)
IS AN ATTACHED SQUARE OF
SAMPLE MASKING MATERIAL
(*ELECTRONICS*, 1966).

as a three-dimensional sample of the material itself, tipped in to the magazine's page for the browser's tactile satisfaction.

**FROM RADIO TO COMPUTING** *Fig. 5.6*, a woodcut print by J. M. Barton, promotes a new magazine, *Space/Aeronautics*. This advertisement was published in *Industrial Marketing*, the meta-magazine of industrial advertising. *IM* was the place where trade advertisements and trade magazines themselves were promoted, critiqued, and analyzed across the entire North American industrial scene. Barton's woodcut features free-floating individual transistors (upper left) as well as imagery suggesting prefabricated circuitry (center). The advertisement was specially printed on a thick, matte-finish paper, and was perforated so that the artwork could be torn out of the magazine and saved. At the bottom of the page were instructions for ordering additional copies of the artwork. Barton's woodcut (an artistic medium that relies on a craft bench rather than an easel) invokes the persistence of craft-based techniques well into the age of automation. From a contemporary perspective, in light of Mumford's historical observation and the resurgence of home crafts, Barton's woodcut appears to contribute to a broader reassertion of the role of craft in the long-term negotiation between human and machine.

The announcement of the transistor in 1948 accelerated the production of printed circuits. Transistors were so much smaller than vacuum tubes that hand-wiring techniques developed for them had to be reconceived, a challenge addressed by the nearly simultaneous emergence of prefab circuitry. The Army Signal Corps Laboratory published its "Auto-Sembly" printed circuit technique in 1952, effectively transferring the technological know-how to the civilian sector.[10] The technique had been developed in part by Bell Laboratories under contract to the army, so even in this case, as elsewhere, the boundary between civil and military research sectors was a partition between classified knowledge and public knowledge rather than a physical boundary between laboratories.

"Auto-Sembly" of prefabricated circuits commenced a series of changes

*Fig. 5.6:*
J. M. BARTON FOR *SPACE/
AERONAUTICS* MAGAZINE
(*INDUSTRIAL MARKETING*,
1960).

"ELECTRONICS" An original woodcut especially conceived for Space/Aeronautics by *JmBarton*
(A full-size, color reproduction without advertising copy, suitable for framing, is available without charge.)

in electronic technology that reciprocally accelerated other processes of automation. The original motivation for developing an automated production technique grew out of the fast-moving market for radios. But when prefabricated circuits met the transistor, in 1952, the combined effect facilitated to an even greater extent the development of faster computers. The result

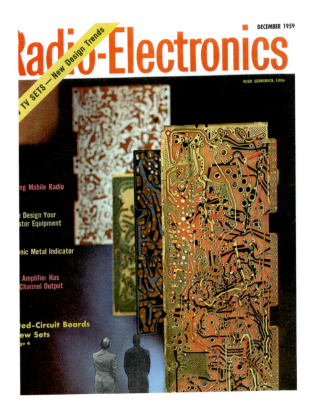

■ DECEMBER 1959

HUGO GERNSBACK, Editor

Radio-Electronics — New Design Trends

TV SETS — New Design Trends

ng Mobile Radio

Design Your
stor Equipment

nic Metal Indicator

Amplifier Has
Channel Output

ed-Circuit Boards
ew Sets
age 4

*Fig. 5.7:*
*RADIO-ELECTRONICS,*
DECEMBER 1959: ON THE
CUSP OF A NEW DECADE,
TWO PEOPLE CONTEMPLATE
A SERIES OF CIRCUIT
BOARDS. ARE THEY ART?
*RADIO-ELECTRONICS* WAS A
MAINSTREAM TECH MAGAZINE
WITH A DIFFERENCE: IT WAS
FOUNDED AND EDITED BY HUGO
GERNSBACK, THE PIONEERING
PUBLISHER AND PROMOTER OF
SCIENCE FICTION LITERATURE.
GERNSBACK'S EDITORIAL
INFLUENCE MEANT THAT
THE INVISIBLE FORCE FIELD
THAT THE FUTURE EXERTS
ON THE PRESENT WAS GIVEN
A FULLER PAGE, AND FULLER
CREDIT, THAN IN COMPETING
MAGAZINES OF THE DAY.

would shift the center of gravity within the electronics industries. Whereas radios had been the site of cutting edge technology at the end of the war, by the mid-1950s the locus of innovation had moved to the realm of computing. "First-generation" vacuum-tube computers began to yield to transistorized computers. It was a shift from those devices that extended particular human senses to the one great device that aimed to mimic the human mind. At this level of technological change, it was not merely a change in physical scale, but a change in the nature of the device. Computers became faster, more efficient machines because the density of transistors enabled a dramatic increase in the complexity of problems that computers could process, and the speed with which they processed them. Transistors accelerated the parallel transition from analog to digital modes of computation taking place within electromechanical computers at the time.

The story of circuits from 1948 to 1962 is one of the unfolding emergence of the board itself as a new electronic component, the process begun by Sargrove. The engineers who first conceived of a circuit board as a potentially indivisible unit could not have dreamt of the advances in semiconductor technology that would follow, yet integrated circuits were developed a mere fourteen years later. In 1959 the first printed circuits were applied to television receiver set design. Of the tantalizingly large and colorful boards on the cover of *Radio-Electronics* (*fig. 5.7*), the front board was manufactured by Motorola. It was the first circuit board to have conductors on both sides of the board. The second board, in blue, was by Philco, while the third was by RCA. The rear board, in red and white, was by Emerson Electronics. Each represented their manufacturer's best step forward in the direction of miniaturization-enhanced television design.

## CIRCUIT BOARDS AND ABSTRACT ART

The angular lines of circuit boards made them a natural fit for graphic artists working in geometric forms. Circuit boards were also the first electronic component to have a visual appearance that was almost wholly abstracted from its function. Electrons have always been invisible to the naked eye and therefore the entire project of representing electronic technologies has always demanded the visualization of an intangible. But the components that populated the first half of the twentieth century—vacuum tubes, CRTs, and even individual transistors and other semiconductors—look different from one another. It is possible to read a three-dimensional hand-wired circuit array like a graphic map: each component has its own look and shape, and when wired together the pattern of their arrangement forms a visual explanation of how the device works.

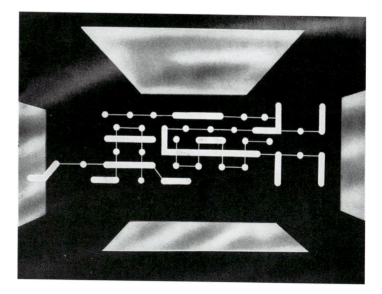

In fabricated circuit boards the connective tissue of circuits was progressively absorbed and subsumed within the grooves and channels of the board's design. Independent of its craft-bench origins, the circuit board also became emblematic of the abstract, untouchable nature of the inner workings of electronic devices. As they developed, the shape and appearance of circuit boards converged toward the two-dimensional drawings on which they were based. The gap between plan and execution narrowed, as circuits became progressively planar and angular with

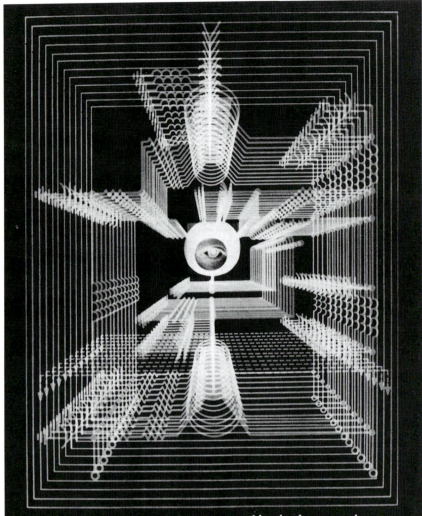

*Look into the vehicles of opportunity at Lockheed...for circuit designers.*

Opportunity broad enough to interest every circuit designer, that's the sweep of electronics assignments at Lockheed. Big, wide-ranging programs that extend from deep sea to deep space. And with ever-growing commitments comes an increasing need for new concepts and major technical advances in flight controls, communications, antennas and state-of-the-art electronics checkout equipment in both spacecraft and fleet ballistic missiles. In addition to its major vehicle programs...Agena, Poseidon, and Polaris, Lockheed is involved in deep submersibles; unique advanced land vehicles; information systems for states and hospitals; and many other technically alluring programs. For complete information, write Mr. R. C. Birdsall, Professional Placement Manager, P.O. Box 504, Sunnyvale, California. Lockheed is an equal opportunity employer.

**LOCKHEED**
MISSILES & SPACE COMPANY
A GROUP DIVISION OF LOCKHEED AIRCRAFT CORPORATION

the shift from tubes to transistors. The circuit board was therefore a ready graphic subject of the mid-century Modern moment in which it emerged. It was already on its own course toward abstraction. Many artists who visually interpreted the circuit board drew on its rectangular frame and irregular lines to explore the link between two kinds of abstraction, technological and visual. In *fig. 5.8* a floating abstract's dimensionality belies the board's flat nature. It has been reduced to the minimum recognizable elements of its matrix nature: the four-sided frame, and a subtly linked circuit network of connection points.

The circuit board artwork created by Willi Baum for the Martin Company (*fig. 5.9*) draws the same matrix motif toward a different outcome. In this work, the rectangle is reconceived as a verdant greensward, with playful yellows, reds, and whites giving the graphic map a look more akin to a parkland maze or suburban rail system than a typical circuit board. Baum's painting was created for a glossy promotional booklet designed to recruit top-ranking engineers to Martin's Denver, Colorado, missile factory, where the company, known today as Lockheed Martin, was at work on

Figs. 5.11 (above) and 5.12 (below): IN THESE TWO ASSOCIATED ADVERTISEMENTS, GENERAL ELECTRIC PROMOTES THE ROLE OF ITS CHEMISTRY LABORATORY IN THE DEVELOPMENT OF PLASTICS AND ALLOYS FOR CIRCUIT BOARDS, AND IN TURN THE ROLE OF THOSE CIRCUIT BOARDS IN FORMING THE "NERVOUS SYSTEM" (COMMUNICATIONS AND NAVIGATION SYSTEMS) OF THE ATLAS ROCKET (*BUSINESS WEEK*, 1959 AND 1958).

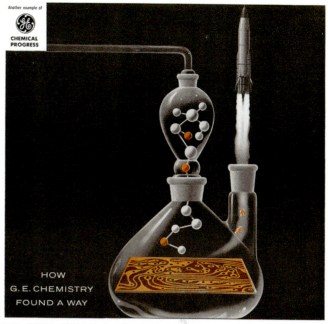

TO GIVE ATLAS A TROUBLE-FREE NERVOUS SYSTEM

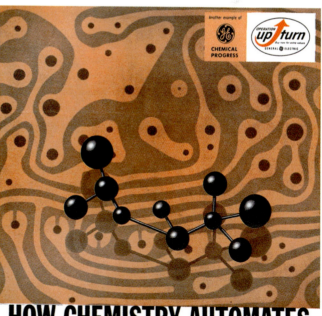

HOW CHEMISTRY AUTOMATES PRINTED CIRCUIT ASSEMBLIES

a commission from the army to build the Titan, a launch vehicle powered by chemical combustion. Titan was a truly massive project, one of the U.S.'s largest rockets, and in the geopolitical environment of the peak Cold War years it was a very important one as well. That era's economic boom encompassed both the Red Scare years of the 1950s and the space race of the 1960s, and as a result government contract firms such as Martin often had swollen hiring budgets. Martin, among many firms, put part of that hiring budget toward commissioning top-notch modernist artists to promote their corporate lifestyle to potential engineers. The Swiss-born and -educated Baum was a good fit for promoting Martin's Colorado environment. A refugee from Europe, Baum sought the Rocky Mountains as a new home in the mid-1950s because they reminded him of his beloved Alps. His early training in Swiss graphic design gave him a depth of skill in working with abstract forms and ideas that served him well during his years creating promotional literature for Martin (until he realized that the Titan was designed to deliver a nuclear warhead, at which point the deeply pacifist Baum promptly quit).[11]

*Fig. 5.10*, from 1967, extends the circuit board deep into Cold War-era graphic concepts. The surveillance trade that dominated that war relied on every type of electronic component working together in a complex system. This image dates to the satellite era, when the flights of robotic espionage spacecraft were coordinated by advanced computing systems. Graphic imagery of electronic networks, as here, became associated with the Argus-like nature of the new technological world. Lockheed was a leading technician in the field of spy satellites even before its 1994 merger with Martin-Marietta (as Martin had become in the interim). The artistic style of *fig. 5.10* is identifiably mid-1960s just as much as the espionage theme, as the integration of photographic elements with graphic elements was relatively new at the time. Here the (photographic) human eye at the center of the network positions the network as an extension of human visual perception, even as the multilevel circuit suggests a network effect of total sensory extension.

**CIRCUIT BOARDS AND SPACE TECHNOLOGY**

The Titan missile was passive as a weapon, an anchor in the arsenal of threats that the global powers maintained against one another during the Cold War. Its ultimate significance was situated not with its deployment in wartime, but in its civil scientific applications. Coincident with the buildup in missile hardware of which the Titan was a part, the U.S. was engaged in a fierce competition with the USSR for accomplishments in civil science. The Titan missile's only "fights" were as a rocket—a weaponless launch vehicle. It famously launched all the Gemini missions that comprised the U.S.'s first two-person human spaceflight program. Similarly, space technology offered a wide arena for intense scrumming between companies over contract dollars and skilled engineers. The space race was more glamorous than its shadowy twin, the Cold War. As a result, many military contractors sought to gain civil contracts with NASA and promoted themselves as space technology companies to potential recruits (more on this in chapter 8). In the Budd illustration that opens this chapter, for example, the ad copy promotes the company's availability for missile and space work.

Along with the shift in the center of gravity from radio to computing, the cutting edge of electronics also moved in an upward direction, from airplanes to missiles and rockets, a bit higher up in the sky. In particular, intercontinental and orbital devices did not rely on radios for communication with the ground; instead, they required complex computer-driven telemetry systems for navigation and guidance. When the USSR launched *Sputnik* in October 1957, there was a dramatic increase in the amount of funding made available for space programs and their constituent components, such as circuits for space-bound telemetry systems. Any advertisement that postdates this development, such as *fig. 5.1*, can be better understood with it in mind.

**FLEXIBLE CIRCUITS**   By the early 1960s, the manufacturing of printed circuitry had advanced dramatically beyond the early sprayed plastics of the 1940s. The military and NASA needed circuits that could be crammed into confined spaces aboard missiles and rockets and that would weigh very little, as every ounce cost fuel. By this time, the electronics industry had developed into a community of hundreds of small companies, each vying for market share against the eight biggest electronics firms in the country. In *fig. 5.13*, the International Resistance Company promotes the new product it launched in 1960, flexible circuits. These circuits, once perfected, would enable changes across the scope of large-object electronics, helping to launch satellites and to enable a range of future computer designs.

Their circuits still touched handcraft, as well as outer space, as the designs were created on a drafting board as late as 1966. The designs were then shrunk and printed onto rolls of flexible plastic that were perforated along their sides in order to move through development stages on a system of belts and spools. Up to one foot wide, these long rolls of circuitry resembled enormous film strips as they made their way through the production system.[12] Copper circuits were then printed directly onto the substrate before the cir-

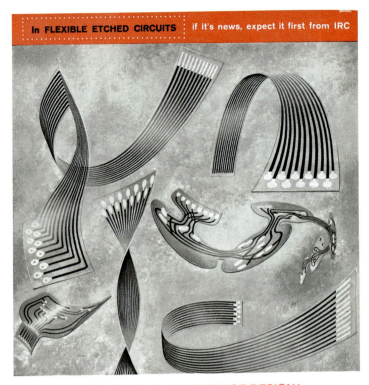

3 DIMENSIONAL· FREEDOM OF DESIGN!

*Fig. 5.13:*
INTERNATIONAL RESISTANCE
COMPANY (*ELECTRONICS,*
1960).

cuit was finalized.[13] As surrealist as the floating, twisting bands of circuitry appear, this artwork depicts the bendy devices' actual appearance.

### INTEGRATED CIRCUITS

Printed circuit boards added capacity continuously throughout the 1950s. Components became smaller and more smoothly integrated into the planar structure of the board over the course of the decade. Then, starting in 1958 and 1959, the world of circuit boards transformed once again. The force for change was the invention of the integrated circuit (or IC): a set of techniques for embedding all the functions of different components, such as resistance, capacitance, and inductance, within the material structure of the circuit board. Just as the transistor harnessed the crystal structures as a tool to make electrons do work, the integrated circuit applied the same principle to the entire workings of a circuit board. In an IC, all functions, from active transistors to all the supporting components, are etched into a single piece of semiconductor material. The crystal structure within the silicon chip can be either physically shaped to perform a particular function or, more commonly, is manipulated in the growth phase with "doping": the technique of introducing specific impurities into the latticework of a crystal structure to cause it to channel electrons in a particular way.

The concept of the integrated circuit had been described in Europe as early as the 1940s; however, the westward shift in the leading edge of the U.S. electronics industry meant that Texas and California were the two most productive sites of practical development of the idea. At Texas Instruments, in the laboratory that produced the first transistorized radio, Jack Kilby was a new hire in 1958. While the rest of the staff were on vacation that summer he had some open-ended time in the laboratory, and that fall he became the first person in the world to prove, using germanium, that it was possible to create a complete circuit from a single piece of semiconductor material.[14]

Over in Silicon Valley, at Fairchild Semiconductor, Robert Noyce was working toward the same end and proved the same idea a few months later, this time using silicon as the semiconductor. Noyce had started his career at Bell Laboratories under William Shockley, just as John Bardeen and Walter Brattain had done, then had followed Shockley to Palo Alto. Along with Gordon Moore and six other engineers (dubbed the "traitorous eight"), Noyce fled the dominion of the erratic and unstable Shockley to form a laboratory at Fairchild.[15] Fairchild's first contract was with the air force in 1957, the conversion of the flight electronics in the B-52 bomber from vacuum-tube to transistor-driven. As Noyce's silicon ultimately proved to be a more effective base material than germanium, it was his integrated circuit rather than Kilby's that provided the template for the future of the IC industry.[16]

In developing their lab techniques for photolithography of integrated circuits, Fairchild first turned to existing home movie technology, using 8mm camera lenses before eventually developing customized tools for the work. Fairchild also engaged in some of the more serious legacy crimes of westward expansion, building a factory in the Navajo reservation in 1965 and employing members of the Navajo nation to build prefabricated circuits.[17] Fairchild promoted the Navajo factory by emphasizing the aesthetic harmony—even the coincidences in basic design philosophy—between integrated circuits and Native American textile design patterns.[18] However, Fairchild found that the Navajo people became open to unionization when confronted with the realities of factory work.[19] When, in 1975, members of the intertribal American Indian Movement cast the group's might behind a unionization effort, Fairchild closed the plant and relocated its circuit manufacturing offshore.[20] A speculative history might explore those resonances between Navajo design and circuit design, imagining how history might have unfolded differently, in favor of labor and handcraft. Instead, the electronics industry contributed to the offshore pursuit of cheap, pliable labor.[21]

Kilby and Noyce, both independently and as affiliated colleagues, faced "tremendous criticism" over the integrated circuit from other engineers within the electronics field as the device produced something of a whiplash effect across industry. In fewer than ten years, electronics manufacturers had made enormous investments in changing the infrastructure of their production and the form of their products to accommodate the transistor as it replaced the vacuum tube. As Kilby remarked in his 2000 speech to the Nobel Committee upon accepting the prize in physics,[22] there were a number of objections to IC technology. Some were practical, such as concerns that semiconductors were not the best materials for making transistors. Others were expressions of worry about the impact of change itself. People worried that if ICs could be made to work, then circuit designers would lose their jobs. As in the case of the concern that television would "kill radio" (see chapter 2), technological change often stirs fears of loss. Sometimes that concern is localized within an industry, as with radio or circuit design. Other times, as with modern concerns that e-books will displace analog reading practices, those concerns unfold at a societal level. In this instance, the concerns were tied in with the broader technological context of the shift from analog to digital modes of computation, a process not obvious to the general public. Ultimately, instead of becoming displaced, circuit designers adapted to miniaturization by becoming more highly trained, specialized, and highly paid than ever before. And as IC-driven minia-

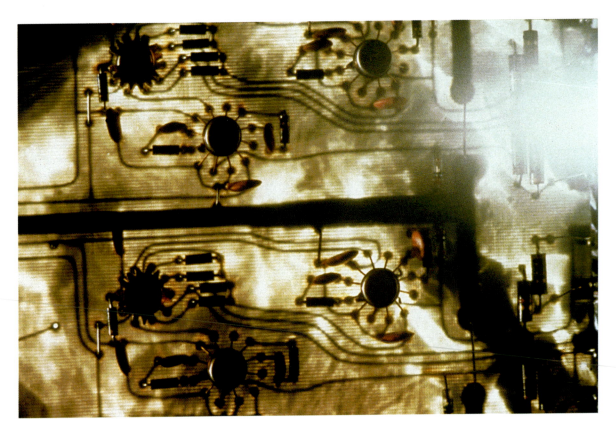

*Fig. 5.14:*
VOLTAGE-CONTROLLED SINE-WAVE GENERATOR. IN 1968 THE ARTIST-ENGINEERS OF PULSA DEVELOPED A SOUND SYNTHESIS AND LIGHTING CONTROL SYSTEM THAT COULD BE MANIPULATED BY SOMEONE UNFAMILIAR WITH ELECTRONICS. WITH EXTENSIVE USE OF INTEGRATED CIRCUITS, THE DEVICE WAS DEVELOPED TO RESPOND TO "RECENT INTEREST IN ENVIRONMENTAL ART." BY PUBLISHING THE COMPLETE DIAGRAMS FOR THE DEVICE IN *IEEE TRANSACTIONS ON AUDIO AND ELECTROACOUSTICS*, THE MEMBERS OF PULSA DONATED THE DEVICE TO THE PUBLIC DOMAIN.

turization of circuits swept the industry in the 1960s, more circuit designers were needed than ever before.

Integrated circuits were a new visual trope, and they facilitated yet one more scalar shift to smaller devices. Yet they were still an extension of the original concept of a circuit board; there is a direct progression from Sargrove's initial formulation of a circuit board as a unified component to the revolution in integrated circuits of the 1960s. ICs were square or rectangular, like circuit boards, and carried the defining flatness of the early circuit board to its ultimate extension. The visible pattern on the circuit of the microchip became a more complex array of grooves, channels, and impressions than had appeared before. These could no longer be read like the graphic map of a fully wired 3-D circuit board, but instead suggested an overflight view of a teeming city with transistors packed as densely as people in buses on microscopic "streets" below.

Circuit diagrams inspired artists as early as the 1920s, when the Japanese artist Tomoyoshi Murarama drew figures with electrical schematics representing the architecture of the human body, schematics that connected the body outward to the newly electrified world. Artists of the 1960s subsequently found in electronic circuits an analog to landscape as well as to the human experience.

Reading a circuit diagram like a map, the Yale-based artists group Pulsa, founded in 1966, used electronic networks to illuminate entire physical landscapes as enormous circuits.[23] For Pulsa the circuit board was a useful tool for investigating unexplored relationships between landscape, technology, and society. Their work drew circuit boards into the realm of art at the same time that modernism began to yield ground to the forces of "dematerialization," a term used by the cultural critic Lucy Lippard to describe the late 1960s' turn toward conceptual art.[24] Just as electronic circuits gradually disappeared into intangibility with the shrinking of circuits, art of the 1960s escaped the confines of material objects. The electronic art that unfolded in the 1970s reliably drew upon the rhyming between microscopic circuits and the intangibility of networked life.

**MICRO-ELECTRONICS** Integrated circuits inaugurated the era of microelectronics. As silicon-based integrated circuits developed, the circuit board became the chip board; then, eventually, the microchip. This progression was a downward-trending shift in scale that paralleled, but opposed, the expanding outward-trending shift in geographic scale that electronics afforded. Circuits became microscopic. Circuit boards yielded to microchips weighing merely milligrams. At the same time, the smaller circuits became, the further the reach of the networks they enabled. Eventually, microelectronics made space travel possible, a subject that is the focus of chapter 8. In the early 1960s, the dramatic transformation of electronics themselves was still the focus.

Gordon Moore closely considered the process of miniaturization that the IC had accelerated. In 1965 he published an article in *Electronics*, written as an observational editorial but now remembered as Moore's Law.[25] This "law" famously posits that integrated circuits would double the number of components they could accommodate every year, as technological advances enable increasing miniaturization and a continually improving balance between cost and production. In 2013 that "law" had only recently begun to break—an incredible forty-eight-year stretch of prognostication by Moore.

Production techniques for integrated circuits developed in the 1960s moved electronic manufacturing away from nearly all the craft traditions and

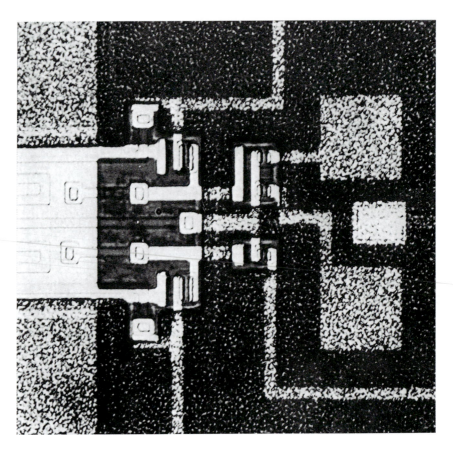

*Fig. 5.15:*
JACQUELINE CASEY FOR THE
LINCOLN LABORATORY (*PROC
IEEE*, 1975).

into bright white, clean rooms. Yet one particular link to craft traditions survived the leap to microelectronics: the very old connection between electronics and weaving. An association between electronics and the craft tradition extends all the way from the eighteenth-century weavers and the development of punched cards (explored in the next chapter) to the 1960s' practice of encoding information into the weave of wiring patterns. In the mid-1950s Japanese engineers at Toko, Inc. developed a method of encoding digital computer memory in the weave of a wiring array.[26] At the same time, braiding (or weaving) was used in England with coaxial cables for stabilizing their electronic capacities, such as resistance and attenuation.[27] Ultimately, as microelectronic circuits were developed they retained the shape of the fine wire leads that are an underlying part of their architecture. In *fig. 5.15*, MIT's graphic designer Jacqueline Casey has stylized a microphotograph of a 1975-era polysilicon integrated circuit, a product of the electronics research wing of MIT, Lincoln Laboratories. In her careful casting of the IC, she lets the circuit's prevalent pattern of squares and right angles assume an artistic mode; to the naïve eye it resembles an etched weaving.

All of these characteristics allowed microcircuits to retain a legacy association with textile crafts, an association that has been encouraged in the intervening decades by the networked nature of the electronic systems that now structure everyday life. The twenty-first-century textile artists Ligorano/Reese, writing about their electronic tapestries woven of fiber-optic thread, explain their approach to integrating electronics with weaving: "Weaving is a social activity. It is about threading narratives and mythology, even language and accounting, with quipoos. Weaving is a shared tradition common to cultures throughout the world in the same way that computers and networks are and have flattened the world, making communication and exchange more common."[28]

The net result of a mere decade's worth of change in circuit technology yielded two major realignments within the electronics industry. First was the lateral shift from radios to computers. Second was a vertical shift, from ground-based technologies to space-based technologies. Next we look at how the information systems within these rapidly evolving systems changed along with the systems themselves, and how that information was represented along the way.

## CHAPTER SIX

# AUTOMATIC AND DIGITAL: THE EMERGENCE OF COMPUTING

It may have been not altogether an accident that the telegraph was invented, not by a physicist . . . but by an artist. The genius of a great artist consists less in creating new things than in combining old and familiar things in new relations. . . . It may have been the artist in Morse that led him to see, in what other men had produced in the field of electrical research, the separate elements which might be synthesized into the telegraph; that enabled him to visualize them in new relationships.

—*Bell Telephone Magazine,* 1944, the centenary
of the invention of Morse code[1]

The rapid development from circuit board to microchip was among a suite of processes that shaped computing technology in its formative years. The addition of circuits to preexisting systems in the fields of data processing and telecommunications propelled the development of computing, along with the ongoing integration of circuits into (electromechanical) mathematical computation systems. Each of these three fields—data processing, telecommunications, and mathematical computing—had been sped up and made more reliable with automatic systems during the century that preceded the widespread emergence of computers. Each field also developed by breaking information into discrete pieces and processing those pieces toward a desired outcome—in other words, becoming digital.[2]

When elements of the three fields were combined with circuits and with one another, the result was the automatic, all-electronic, all-digital computer.

The respective features of these fields yielded a new outcome more powerful than any one of them: computing, a new industry of the 1950s. Over the course of that decade computing would come to dominate the field of electronics, taking over its vanguard before defining its own universe of influence.

Artists responded to this environment by cultivating the graphic evidence of these processes. Across automation and "digitalization"[4]—the change in a system from analog to digital technology—the common type of mark left by the different systems was a bead, a dot, or a hole in a card or strip of paper. These were both literal manifestations of machine processes, such as the hole in a punched card, and symbolic representations, such as the use of abacuses to signify mathematical calculation. As these elements came to signify the emergence of networks and digital systems, the artwork based on them drew on their abstract and particulate nature to build associations with these developments as well as with their procedural origins. They were often used in combination, as in *figs. 6.1* and *6.2*, as the nascent computer industry of the 1950s sought to convey its multiple origins and many capacities. The result is a set of images that harness the geometric simplicity of these motifs and animate them to convey a range of developments in the computing field.

Walter Murch's soft-focus work in vivid oils (*fig. 6.1*) exemplifies these strategies. This 1959 composition for First National City Bank of New York neatly combines several of the motifs that artists developed to depict the decade's swift incorporation of tech-nologies of automation and "digitali-zation." A yellow punched card floats in front of strips of paper tape, while a large black abacus dominates the painting. Between these elements, three constituent technologies that combined to form the computing industry are represented, with lines connecting points on the globe sug-gesting international telecommunica-tions and punched cards representing data processing. *Fig. 6.2*, a two-color graphic from the cover of the maga-zine *Data Processing for Management*, uses most of the same motifs to pro-mote an automated audiovisual cur-riculum for management.

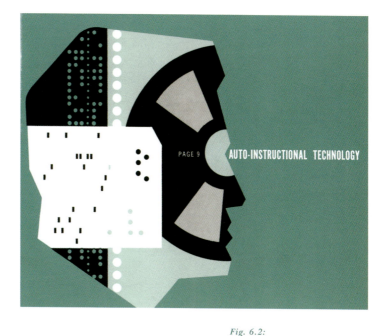

Fig. 6.2:
*DATA PROCESSING FOR MANAGEMENT* MAGAZINE COVER. MECHANICAL MAN: A PUNCHED CARD, A STRIP OF FIVE-BIT PAPER TAPE, A STRIP OF FILM, AND A REEL OF MAGNETIC TAPE MAKE A FOUR-PART SYSTEM (1964).

**INDUSTRIAL AUTOMATION**   Automation in its broadest sense refers to the sum of mechanical techniques developed since the eighteenth century to introduce feedback, or self-correction, into industrial systems. As it contributes to the development of the computer indus-try, automation refers to the application of such techniques first to data pro-cessing and subsequently to mathematical calculation; later to the entire field. Among the most visible of those applications is the use of punched cards and paper tape as input/output devices; less visible are the switch systems devel-oped to control telephone exchanges and streetcar movements.

The switching systems that operated many Midwestern municipal rail services at the turn of the twentieth century were electromechanical con-trol systems built by Westinghouse (*fig. 6.3* is an example of their artwork). Best known in its early days for electrical networks, the company used its toolbox to build power and transportation systems, as well as the first indus-

**Westinghouse**

trial robots (more about them in chapter 9). The company developed its first "supervisory" control system in 1924. This system, called Televox, was a vacuum tube–driven, binary logic system, a switch-based decision tree allowing for automatic control of electrical substations.[5] With their on/off switch-based binary logic systems, control systems could be "programmed" to make decisions based on contingencies within a system. Systems like Televox were therefore rudimentary forms of artificial intelligence (AI). Together with other switching systems such as the telephone network switches developed at Bell Laboratories, these systems modeled on/off decision trees that contributed to the development of binary computer logic. They were antecedent to the binary "flip-flop" circuit design that formed the logic gates and information storage systems of early computers.

Westinghouse was a government contractor for war production in World War I, a tradition the company maintained throughout its existence. With this kind of experience prior to 1930, Westinghouse rolled with the technological changes of the coming decades. By the time the satellite era arrived the company was providing essential infrastructure support and control systems, including computing systems, to NASA as well as to the military. Its industrial footprint in the mid-twentieth century was as large as that of its rival General Electric, but, like its former employee Nikola Tesla, the company left behind fewer traces at the intersection of art and industry than many of its contemporaries.

*Fig. 6.3:*
WESTINGHOUSE. THE
COMPANY'S LOGO (UPPER
RIGHT) WAS CREATED BY PAUL
RAND (*ELECTRONICS*, 1961).

### ABACUSES: THE EMERGENCE OF DIGITAL SYSTEMS

Electromechanical mathematical computation was developed during World War II first to assist with ballistic trajectory calculations and later to work the vast simultaneous differential equations that had to be solved in order to predict the detonation of an atomic bomb. In the course of these army- and navy-led projects, mathematicians and engineers developed a binary mode of automatic computation—integrating automatic switching systems—that was also digital. When the computer engineers J. Presper Eckert and John Mauchly boosted the digital computational logic system they were building for the army[6] with the electronic power of over seventeen thousand vacuum tubes, the computer that resulted, the ENIAC, became, exponentially, the most powerful computational tool ever built.

Like electricity, information can be transmitted as either wave or particle, through either analog or digital systems.[7] The use of pebbles as counting tools millennia ago are the first use of "particles" in the service of computation (the word "calculus" is Latin for "small stone"). The abacus, a computational tool used throughout recorded history, is an antecedent of digital computation. It allows numbers to be formulated and calculated using a codified, symbolic system of beads. The terms *digital* and *analog* refer to whether the data being processed has been encoded into a simple and irreducible system of representations, or whether it is being roughly conveyed, perhaps even in literal terms.

An abacus is digital in the most elementary sense: it expresses complex numbers and numerical relations through a simple and irreducible symbolic system of beads. The same is true of quipu, a Native American accounting system using knots on rope. A slide rule, by comparison, is analog: numbers and numerical relations are expressed through a spatial relationship that must be interpreted by the person using the device. When the carrying capacity of an analog system is doubled, such as doubling the length and numerical range of a slide rule, the system can process double the amount of information. With a digital system, even an abacus, if the carrying capacity of the system is doubled then the range of possible calculations expands exponentially rather than arithmetically.

The earliest electronic devices made from vacuum tubes, including radios and all cathode-ray tube-based devices, were analog. They were electronic machines, but the smooth flow of a radio or television broadcast was strictly waveform (analog) technology. When both automated systems and digital information processing techniques were combined with electronic circuits, artists developed a visual repertoire to communicate the significance of these dimensions in the field of electronics. Their meta-significance was the emergence of computers as the dominant electronic technology; on the associated pages of trade magazines, the more quotidian task was to introduce computers and, in doing so, convey the significance of their different systems. The resulting graphics emphasize the particulate nature of digital information, and the mechanical hardware—and eventually software—of computers.

The abacus in particular conveys the idea of mathematical calculation in addition to its beads suggesting a digital system. It is a charismatic, historic object that is capable of suggesting the future of computation as well as the past. With a legacy that reaches back to the China of millennia ago, the abacus was well positioned to suggest a portal between past and future, with its nature as a manual tool also suggesting human touch (as in *fig. 6.4*). Its rectangular shape forms a gate at the boundary between the touchable world and the abstract, mechanized world of information.

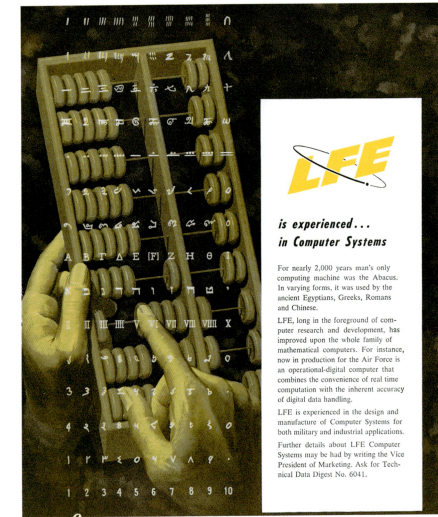

These characteristics made it a useful tool in the process of naturalizing new technologies, and it was widely deployed in advertising to promote new computer systems. It helpfully demodernizes the idea of a digital future by suggesting that "digital" is as straightforward as a composed rack of wooden beads.[8]

In 1947 the eminent graphic designer Paul Rand created a photogram (a camera-less photographic image) of abacus beads to decorate the cover of his book *Thoughts on Design* (*fig. 6.7*). Rand would go on to create the corporate logos for IBM and Westinghouse, among many others, being well-positioned for these jobs because of his early adept fusion of design with technology. His abacus design implied that the qualities that that device represented could be equated with postwar changes in design and representation.

The forces of automation and "digitalization" were independent of each other, traveling like parallel circuits on offset chronological trajectories for centuries until they were combined. The 1940s was the laboratory decade for the development of these processes, while the 1950s saw them publicly inte-

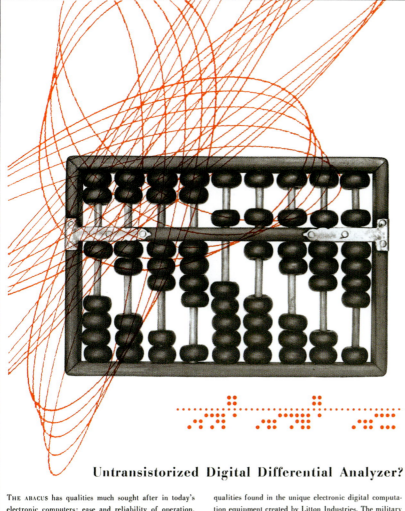

## Untransistorized Digital Differential Analyzer?

THE ABACUS has qualities much sought after in today's electronic computers: ease and reliability of operation, low investment, and minimal maintenance. These are qualities found in the unique electronic digital computation equipment created by Litton Industries. The military and industrial applications for this equipment are many.

**LITTON INDUSTRIES** BEVERLY HILLS, CALIFORNIA
*Plants and Laboratories in California, Maryland, Indiana and New York*

DIGITAL COMPUTERS AND CONTROLS    RADAR AND COUNTERMEASURES    INERTIAL GUIDANCE    MICROWAVE POWER TUBES
PRECISION COMPONENTS    AUTOMATIC DATA PROCESSING SYSTEMS    SERVOMECHANISMS    SPACE SIMULATION RESEARCH

*Fig. 6.5:*
IN 1956 COMPANIES PRODUCING VACUUM-TUBE COMPUTERS WERE PROMOTING THEM AS "UNTRANSISTORIZED" (AS IF THAT WERE AN ADVANTAGE), EMPHASIZING THAT ADVANCED DIGITAL SYSTEMS WERE NOT NECESSARILY RELIANT ON TRANSISTORIZED CIRCUITS. LITTON INDUSTRIES WAS A STORIED ELECTRONICS FIRM; ITS OFFICIAL CORPORATE HISTORY SHAMELESSLY REPORTS A DESPICABLE CORPORATE CULTURE RIFE WITH OVERT RACISM AND DUPLICITY. LITTON WAS ALSO UNIQUE IN AMERICAN INDUSTRY IN REACHING OUT DIRECTLY DURING THE COLD WAR TO BOTH THE USSR AND TO COMMUNIST CHINA TO EXPLORE POTENTIAL UNSANCTIONED TRADE AGREEMENTS. THE COMPANY PUBLISHED ITS CORPORATE REPORTS IN RUSSIAN AND DIRECT-MAILED THEM TO SOVIET INDUSTRIALISTS, TEASING THEM ABOUT POTENTIAL "OPPORTUNITIES" (*PROC IRE*, 1956).

*Fig. 6.6:*

LOCKHEED. THE AD COPY USES A NEUROLOGICAL METAPHOR TO DESCRIBE THE COMPANY'S COMPUTER RESEARCH, BUT THEN EXPLAINS THE LINK BETWEEN DESIGN AND TECHNOLOGY: "PROGRESS IN THIS FIELD IS SYMBOLIZED BY THE ABACUS—EARLIEST FORM OF COMPUTER." TO THE CONTEMPORARY EYE THE TOTAL COMPOSITE IMAGE SYMBOLIZES SOMETHING MORE: THE RIDICULOUSNESS OF THE PERSISTENT LITIGATION BETWEEN TECHNOLOGY FIRMS OVER RIGHTS TO BOTH THE CONCEPT AND THE LOOK OF TABLET DEVICES. THE CONCEPT FOR TABLET DEVICES, AND THEIR LOOK, PREDATE THE CONTRIBUTIONS OF ANY PARTY TO THE VARIOUS TWENTY-FIRST-CENTURY LAWSUITS THAT CONTEST THAT QUESTION. NOTE THE USE OF A SIX-BIT PAPER TAPE DOT PATTERN AS A DECORATIVE DESIGN ELEMENT (*AVIATION WEEK*, 1959).

grated into computers in the context of business and industry. If any one of these elements had combined differently, we might be living in a world dominated by transistorized analog computers or by tube-driven digital computers, or even, as explored in the Bruce Sterling and William Gibson novel *The Differ-*

*ence Engine*, mechanical digital mainframes. In the 1950s these "new" component parts began to be introduced and naturalized through commercial artwork to the same communities of workers, engineers, and businesspeople who were simultaneously being oriented to vacuum tubes, circuit boards, and transistors. By 1960 no one in the electronics field could afford to be uninformed about automated data processing systems any more than the basics of computer hardware.

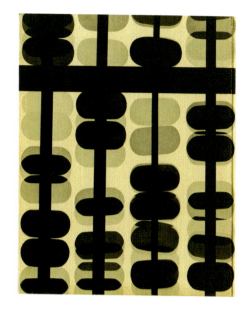

*Fig. 6.7:*
PAUL RAND, *THOUGHTS ON DESIGN* (1947), COVER DETAIL.

**PULSE CODE MODULATION** At Bell Laboratories, during the same years that Eckert and Mauchly were building ENIAC, the engineer and mathematician Claude Shannon and other colleagues were experimenting with digital technologies in telecommunications. It was the height of the radio era, and Shannon's research concerned the problem of how to expand the capacity for voice transmission within a transmission wire without widening the wire. The result was the first digital communications technology, an application of pulse code modulation (PCM), a system of encoding the signal transmitted through telephone wires or over a radio network into a digital signal. The objective was to maximize the carrying capacity of telecommunications systems in the face of muddying factors like resistance or static on the line. When sound was translated into a digital signal system (pulse code), the total amount of information that a given line could carry expanded exponentially.

PCM transformed telephony and radio, and it also broadened the reach of digital systems. As those systems multiplied across technological domains, the graphic motifs associated with them had an increasingly broad field of reference.

**LANGUAGE CHANGES** It was little coincidence that the transistor and PCM were developed in different departments at Bell Laboratories at about the same time. When applied to com-

# A Survey of FACTS, not opinions; made to get information, not to prove something

This is a reproduction of a survey report card, filled out to show the living standards, reading habits and newspaper preference of a family.

T HE investigator carefully recorded on the card the name, address and telephone number of the family, the district in which it lived, the ownership of automobiles and the newspaper reading habits. All questions related to facts, not opinions. No leading questions were used. No leading questions could be used because the investigators did not know the purpose of the survey nor for whom it was being made.

In a survey conducted in this manner, bias is impossible and error is negligible.

## Three Accepted Measures of Standards of Living

No estimates of income or expenditure or any other doubtful figures were employed in this survey. Standards of living were measured, first, by the type of residential district, as determined by the City Plan Commission; second, by the presence of a telephone in the home; third, the ownership of an automobile.

puter systems, digital technology sped up the processing of information so quickly that only binary data could keep up with the processing speed, a situation that also applied to telecommunications in the late 1940s. Binary data systems convert alphanumeric values to coding systems in which each possible value is expressed by a combination of signals that are each either "on" or "off." The numbers and letters conveyed through analog telecommunication systems were commonly known as digits. Shannon's Lab colleague John Tukey proposed that they be renamed when they were turned into binary form: that they be called "bits," a contraction of "binary digit."[9]

In 1959, IBM programmers working on what would be the 7030 Supercomputer developed a novel method of entering 64-bit lines of code into the computer (equivalent to eight 8-bit lines of paper tape). Within that format it became germane to identify "words" within the code, as expanding upon the previous limit of 8-bit length permitted a more widely variable vocabulary. These "words," drawing on the term "bits," were named "bytes" by the IBM software engineers.[10]

**PUNCHED CARDS**    Electromechanical data processing machines were already using digital systems long before they were harnessed to electronic circuits. Mechanical and electromechanical techniques of automated data processing had their roots in very old systems of encoding data in discrete, quantifiable amounts. Eckert and Mauchly's ENIAC used punched cards for programming and the output of results, a technology originally developed in France in the eighteenth century by textile manufacturers, who used cards to "program" silk weaving patterns. In the first decade of the nineteenth century, improvements to punched card systems for looms by Joseph Marie Jacquard enabled this technology to accelerate and to cross borders, marking the emergence of digital automation in manufacturing.

Punched cards are binary programming tools: Programs for weaving or calculation are encoded through patterns of holes in the cards, and as the cards are reusable the programs are stored on them when not in use. For each possible position on the card where a hole might be punched, there either is, or is not, a hole—hence a binary, proto-digital data system. As ordinary as this system sounds, it was in fact the first computer memory technology, as well as the first computer programming tool. Our contemporary references to "stored programs" for computers have their origins in the punched card.

England's craft- and textile-based economy was abruptly interrupted by

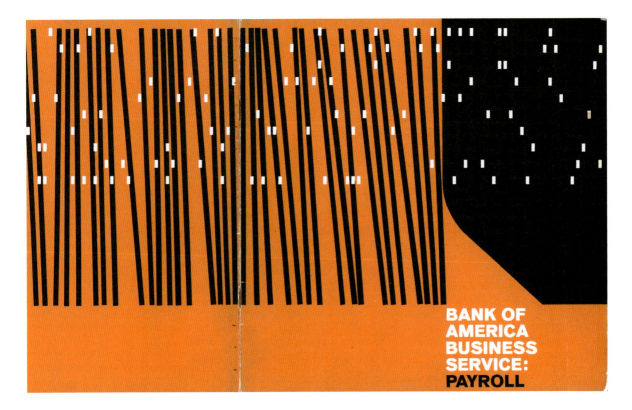

BANK OF
AMERICA
BUSINESS
SERVICE:
PAYROLL

*Fig. 6.9:*
WILLI BAUM FOR BANK OF AMERICA. THIS PUNCHED CARD ARTWORK IS FROM A PAMPHLET PRODUCED TO HELP EMPLOYEES EXPLAIN THE BANK'S NEW COMPUTERIZED PAYROLL SYSTEM. PUNCHED CARDS TYPICALLY HAD A DIAGONAL CUT ON ONE CORNER TO ORIENT THEM CORRECTLY; HERE BAUM HAS HUMANIZED THE CARDS BY INTRODUCING IRREGULARITY TO AN OTHERWISE MECHANICALLY REGULAR DESIGN (1961).

the introduction of punched-card looms in the early nineteenth century. Millions of people lost their jobs, establishing a pattern in which technological development was a counterforce against social justice.[11] Punched-card technology soon migrated to other technological systems. In the 1840s the English inventor Charles Babbage adapted the idea of punched-card programming in designing his "difference engine," his plan for a mechanical computer.

The use of punched cards accelerated after the U.S. government statistician Herman Hollerith adapted them to track social information for the purpose of the census.[12] In 1880 Hollerith noticed that the time it took for census-takers to process their data from that year's decadal survey guaranteed that the next survey would fail if the system were not automated: the population density and complexity of the U.S. was expanding so rapidly that an 1890 survey conducted with 1880 technology would not generate results prior to 1900. In other words, population growth demanded automation of the census; without it, the results would be all but useless. In response, Hollerith developed the first automated data collection machine, programmed with punched cards. The electromechanical "Hollerith machine," as it became known, was running in time to make the 1890 census the most comprehensive and quickly understood census in history. Its impact far exceeded that of any prior census, influencing American politics, social sciences, and cultural geography.[13]

Hollerith machines, and the cards that came to be known as Hollerith cards, are perhaps the longest-standing data processing machines yet deployed by humankind. Punched cards were still in regular use until the 1980s; whether we count back as far as Jacquard looms or merely to Hollerith's 1880s' invention, it is hard to attach a longer timeline to any comparable machine technology. Hollerith formed his own consulting company and—to summarize a long story that is told elsewhere in detail—his company eventually became International Business Machines, or IBM. IBM's postwar dominance in computing is based in part on its history of continuous experimentation with ways to accelerate data processing using electronics. IBM's research laboratory was a

**Announcing UNIVAC 120**

peer to that of Bell Laboratories, and its situational advantage enabled it to become the most powerful company in the U.S. when computing became a dominant force across many industries.

Punched cards were adapted to electronic systems with the use of sensitive metal pins that would drop through the patterned holes, completing a circuit where the hole was punched and "telling" the computer what to do through the resulting circuit patterns. They formed the backbone of input/output systems and of stored-program computing for decades. By 1955 they were helping to automate the assembly of printed circuit boards.[14] Punched cards were steadily adaptable to the forces of digitalization and miniaturization that swept computing through the 1950s, the 1960s, and even the 1970s. Their impact on the culture of the industrialized world cannot be overstated. As broadly applicable tools of automation, they came to permeate practically every domain of life in which data could be crunched. For a great many

*Fig. 6.10:*
REMINGTON RAND, MAKER OF THE FIRST COMMERCIAL VACUUM-TUBE DIGITAL COMPUTER, THE UNIVAC, AND A LONGTIME MANUFACTURER OF PUNCHED-CARD DATA PROCESSING MACHINES, MADE A STRATEGIC MOVE IN 1950 THAT BROUGHT IT ON PAR WITH IBM WHEN IT HIRED ECKERT AND MAUCHLY. (*BUSINESS WEEK*, 1954).

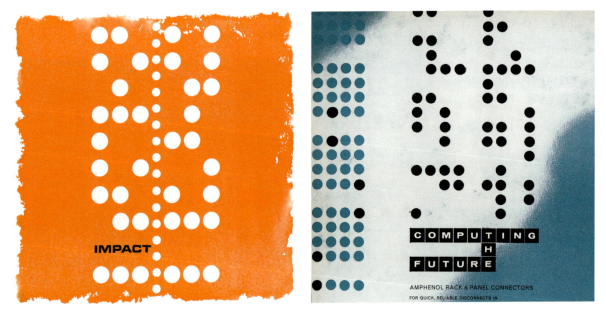

Fig. 6.11:
TUNG-SOL ELECTRIC INC. (*SCIENTIFIC AMERICAN*, 1961).

Fig. 6.12:
AMPHENOL-BORG (*ELECTRONICS*, 1961).

people they also became symbols of unwanted social control: resistance to being treated as a number rather than a person[15] has its origins in the use of punched cards to organize information about people within government, education, and industry.[16] The long-standing criticism of technology as a force for dehumanization has changed over the decades, mostly to adapt to developments in the ways that technology facilitates social control.

Punched card technology was adapted as an artistic medium as early as the 1850s by those who recognized its applicability to the strings of a piano. Punched paper rolls were joined to modified pianos by nineteenth-century inventors, a technology that was seen by some as downgrading the art of piano playing to the status of programmed entertainment. Vertical player pianos, however, brought a version of piano music to many places, especially in the western U.S., where otherwise there would have been none. In the twentieth century radio rendered player pianos obsolete, but in the 1940s they were discovered by one of the century's most original modern composers, Conlon Nancarrow. Less famous than George Antheil, another composer who used player pianos, Nancarrow used punched paper rolls for reasons similar to Hollerith's: he wanted to push his explorations of rhythmic patterns beyond what manual techniques would allow. The precise and abstract arrangements he heard in his mind's ear could not be realized without automation; as such, they exemplify the modernist gesture in art. Eventually Nancarrow had a custom piano-roll-punching machine built. His intense, fast, staccato pieces, collected in the four-volume *Studies for Player Piano*, have since been recognized as essential contributions to modern music.[17]

**PUNCHED CARDS . . . AND PAPER TAPE** Samuel Morse's status as a fine painter is tangential to his role as developer of Morse code. Yet the remark about his work in *Bell Telephone Magazine* that opens this chapter contributes to an understanding of the mutual origins of electronics and fine art. The code that Morse invented in 1844 was a dot-and-dash symbol system that changed human communication. It was a binary, digital system—a code that translates analog messages into patterns of 0 and 1 (dot and dash). Using Morse's code to stand in for alphabetic language, the telegraph system dramatically accelerated the amount of information that could be transmitted through electrical wires. Telegraphy became a reliable and international communication network, one that offered an electrical assist to the nineteenth century's steam-powered processes of globalization. And all because information was digitized: a simple recoding that formed the foundation for many twentieth-century changes in information technology.

Morse code was developed in the U.S. at the same time that Charles Babbage and Ada Lovelace were working in England to adapt the concept of punched cards to the service of mathematical computation. The telegraph system was a technological cousin to the punched card: paper tape developed to feed prepared messages into a telegraph system was inspired by punched-card technology and was subsequently adapted as an input/output technology for computers that was cheaper than punched cards. As a result, Morse code and telegraphy were regular reference points for twentieth-century engineers developing automated information systems. "Instead of having only one telegraph line along which instructions can be sent, ENIAC has more than 100 such lines," wrote Edmund C. Berkeley in his popular 1949 book on computing, *Giant Brains*.[18]

Paper tape, adapted directly from telegraphy to computing, offered external storage and continuous feeds of modest amounts of information or computing instructions. Its primary advantage over punched cards was cost, though it was far more fragile. Over time, the tape was reformulated, for durability, as a paper-coated mylar base; still, the moniker "paper tape" stuck forever. Paper tape punch patterns are organized along a continuous central

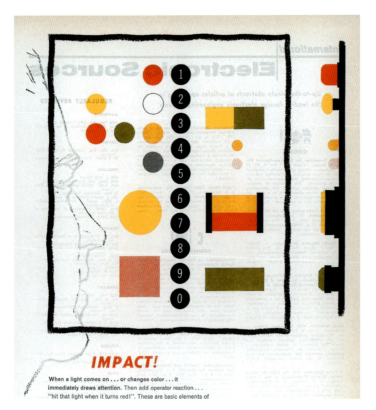

*Fig. 6.13:*
CONTROLS COMPANY OF
AMERICA (*ELECTRONIC
INDUSTRIES*, 1964).

feed line of small holes (as in *fig. 6.11*). When the tape is punched with a code pattern, by a device attached to a typewriter, the central feed line holds the tape on the sprocket. Surrounding the feed line are a set number of channels: either five, six, seven, or eight. With the feed line forming a vertical axis and the line height of each horizontal row established by the distance between the sprocket positions on the feed line, an inch of paper tape could hold up to 72 "frames" for a potential hole punch (or bit).[19] *Fig. 6.11* clearly shows the central feed line of small holes with a total of six frames, three on either side of the central line. *Fig. 6.12* is a more stylized variation: its artwork depicts an 8-bit tape, but the central feed line has been decoratively omitted, adding a layer of abstraction.

The language of paper tape frames was adapted from Shannon's pulse code modulation: a bit is either the presence or absence of a hole within a frame, and each horizontal row of bits conforms to an alphanumeric value in natural (human-readable) language. Several different coding systems were used to link specific bit patterns to specific letters or numbers, and the systems were often proprietary. The market for automatic computers in the 1950s was anchored by heavy industry; large, competitive companies tended to have their own coding systems, a setup guaranteed to frustrate clients trying to use computers as communication tools (how little things have changed). The systems did have in common the look of the visual pattern generated by the coding of natural language into a bit system. The language of data processing dominated the developing computer industry into the 1960s. Systems marketed to business were typically described as electronic data processing systems, and if those systems used mathematical computers within them the feature was mentioned in the details of the promotional literature.[20]

Seven-bit tape code became the foundation of ASCII, the American Standard Code for Information Interchange. Formalized in the mid-1960s by the

American Standards Association, early ASCII was an antecedent to Unicode, the standard cross-platform encoding system for computer-generated text—an antecedent that extended directly from telegraph code.[21] Expanded and redeveloped several times over the next fifty years, ASCII remained the industry standard for computer-to-computer communication until the mid-2000s. Seven-bit ASCII had a blank space at the end of each line, until successive developments pushed it to an 8-bit code. Eight-bit tape became the foundation of calculations about computer memory. Memory systems in early 1980s' personal computers and floppy disks, which were organized in byte numbers in multiples of eight, are a direct legacy of 8-bit tape (64KB/512KB [kilobyte] memory, for example).

The image of punched dots arrayed on paper tape was a remarkably persistent and generative visual trope. After 1954 it appears in many places, both in and out of context, pointing to the emergence of computing as a dominant electronic technology. To a much greater extent than the holes in punched cards, the bits in paper tape escaped their original reference point to become a freestanding design motif. While *figs. 6.11* and *6.12* are relatively representational and promote their companies' products and services in the realm of stored-program computing, *fig. 6.13* is something else entirely: it promotes push-button navigation systems for elevators. Though the system being promoted was an automatic, switch-based electronic system, it's nevertheless a whimsical stroke by the artist to cast elevator buttons as the central feed line within a graphic design that fictively mimics a paper tape array of dots.

Paper tape dots were a surprisingly durable and even portable motif. In *fig. 6.14*, a pattern derived from a 5-bit punch array is mobilized to promote Mallory's line of ever-shrinking transistors. The dot motif has escaped the frame of reference that governed its origins and expresses something else entirely. This makes sense, given that dot patterns relate to digital technologies as widely disparate as the Braille writing system, Morse code, and Morse's computational descendants.

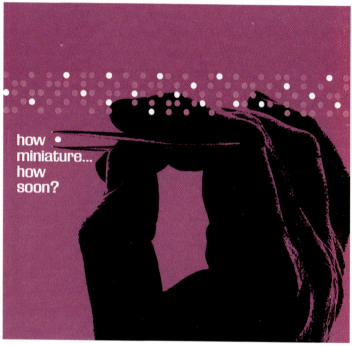

how
miniature...
how
soon?

Fig. 6.14:
P. R. MALLORY & CO.
THE AD COPY PROMOTES
"MICROMINIATURE
ELECTRONICS"—
TRANSISTORS—THEN
INVOKES THE PAPER TAPE
GRAPHIC MOTIF TO CREATE
A CORRESPONDING VISUAL
IMPRESSION (*BUSINESS WEEK*,
1963).

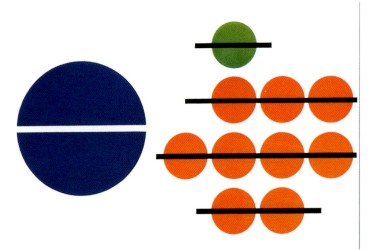

*Journey of a red balloon.* Fig. 6.15 (top left):

UNDERWOOD WAS BEST KNOWN FOR TYPEWRITERS BEFORE BRANCHING INTO ELECTRONIC DIGITAL COMPUTERS IN THE MID-1950S WITH ITS ELECOM SERIES. IN THIS AD ARTWORK, THE ABACUS MOTIF IS ABSTRACTED BUT ALSO STYLIZED TO ASSOCIATE WITH THE COMPANY'S HISTORY IN KEYBOARDS (*BUSINESS WEEK*, 1963).

Fig. 6.16 (top right):

A 1950S' STARTUP, GENERAL TIME SPONSORED THIS AD TO PROMOTE ITS "TRANSACTER" SYSTEM, AN "INSTANTANEOUS" DATA TRANSMISSION SYSTEM. THE RED BIT DOT HAS SHATTERED ITS GRAPHIC PARAMETERS WITH ITS SPEED OF MOTION (*BUSINESS WEEK*, 1960).

Fig. 6.17 (right):

A MORE GRAPHICALLY MATURE STYLIZATION OF PUNCHED DOT CODING, PROMOTING GIANNINI CONTROLS CORPORATION'S "DATEX" ENCODER CODE PATTERNS. IF THE RED BIT DOT BOUNCED DOWN FROM THE ABOVE IMAGES, IT WOULD FALL RIGHT THROUGH THE PUNCHED HOLE (*ELECTRONICS*, 1966)

### MAGNETIC TAPE

The third input/output tape system in wide use during the 1950s and 1960s was magnetic tape (*fig. 6.17*). Introduced by Univac[22] in 1951, it was spooled in reels, then offered to the computer as an input/output technology similar to cards or paper tape. The computer made magnetic marks on the tape to form a pattern conveying information through binary coding.[23] Like its predecessors, magnetic tape had to be read by a secondary device that would translate its bit marks into natural language. Reels of magnetic tape had limited visual charisma, but they turn up as graphic elements in a few places, most commonly in combination with other elements to convey the idea of information processing, as in *figs. 6.2* and *6.18*. Magnetic tape was much more dense and compact than paper-based input/output systems and could hold far more information. This made it particularly well suited to business applications, such as payroll systems, that had a large number of possible solutions to every programmed "problem." Throughout the 1950s and 1960s, trade magazines like *Datamation, Data Processing,* and *Electronics* were peppered with ads for format-to-format conversion devices: paper tape to mag tape, punched cards to paper tape, and every other permutation. As the years passed, magnetic tape was never fully discontinued as a data storage technology. In 2012, after years of decline, its use began to increase again, for all the advantages of speed and reliability that it possesses relative to spinning-disk storage.[24]

**MAINFRAMES** In 1955 mainframe computers gained new traits that expressed the three forces of miniaturization, automation, and digitalization. The IBM 608, introduced that spring, was transistorized and fitted with printed circuit boards. It used more than three thousand transistors and had magnetic core internal memory; punched cards were its primary input/output technology.[25] Results from mathematical calculations could be punched out at twice the rate as the machine's immediate predecessor, a vacuum-tube model. Notwithstanding the advances in miniaturization, computers like the 608, which combined elements of electronic data processing systems and digital computational architecture, remained behemoths. Smaller than the truly massive room-sized, tube-driven experimental machines such as the ENIAC and the Mark series, they were comparable in size to a large car of today.

Their uncomely physicality posed a challenge to the artistic project of visualizing new technologies. The transistor was too small for its morphology to lend itself to graphic stylization; the mainframe computer had the same problem in the other direction. The "outside" of the machine—the mainframe computer itself—was a new device as much in need of naturalization as the components "inside the machine," yet its physicality failed to convey its significance as a new machine. At the same time, graphic artists were challenged by the emergence of photography as an increasingly dominant medium in advertising artwork.

Where art meets technology in this realm of data (or information processing), artists redefined "the machine" in visual terms that referenced its processes and products over its physicality. Extending the practice of depicting the "inside" of the machine to "bits" of data, artists drew on abstraction and the very modern geometric form of the circle. This process echoed the relationship between the transistor and its symbol-based, geometric representations in graphic art. The trend in fine art toward conceptual work har-

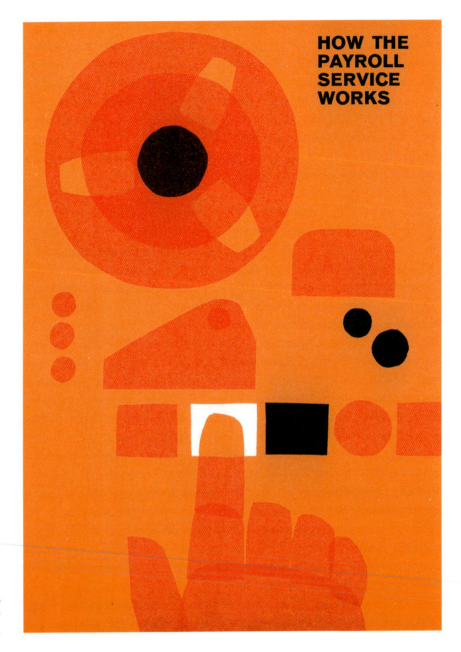

**HOW THE PAYROLL SERVICE WORKS**

*Fig. 6.18:*
WILLI BAUM FOR BANK OF
AMERICA, 1961.

monized with the direction that artists had to take with regard to computers. The result was a range of strategies for depicting *information* as the product of the new machines.

As an abstract concept, the idea of "information" in this context was at least suitable to convey the increase in computer power that resulted from the combinatory inventive processes of the 1950s. In its abstractness, it was also in harmony with the dominant trends of the era in both fine and graphic art. However, the associated trope of the "age of information" was reductionist even as it cursorily explained an important idea to the general public. There was pushback against unquestioned automation even within industry and academia—not only in the cultural sphere. Warning in 1960 that an ascendant crisis in automation was invisible because "electronic controls do not blow up" when they malfunction, the computer scientist George Steele and the academic Paul Kircher argued in the book *The Crisis We Face: Automation and the Cold War* that automated systems were being implemented too quickly.[26] Their work anticipated the politically oriented advocacy group of the 1980s Computer Professionals for Social Responsibility. Five years later, the futurist Alvin Toffler would publish the essay that presaged his book *Future Shock*, promoting a counternarrative of dangerous "information overload."[27] In the context of these and other varied responses to the emerging dominance of automatic computing, the "age of information" moniker appears both totalizing and dismissive. It employs a universal good (information) and symbolically sets it outside the boundaries of political discourse by assigning it the privilege of being its own "age."

Granted that Toffler's counternarrative oversimplified as much as the original idea that it inveighed against, it is nonetheless worth noting that the graphic language of industrial advertising was addressing a subset of the general public that would have been exposed to a very mainstream yet complicated notion of the "age of information." In this environment the images in this chapter develop graphic strategies that touch on both alienation (the floating, disconnected dots) and the simultaneous positive connotations of information processing power that their sponsors surely sought to guarantee by association. These have all been abstract or at least nonlinguistic images; the next chapter looks at the cultivation of typographic art for the same purposes, with a surprisingly different result.

## CHAPTER SEVEN

# VISIBLE LANGUAGE

3.14159 26535 89793 23846 26433 83279 50288 41917 69399
37510 58209 74944 59230 78164 06286 20899 86280 34825
34211 70679 82148 08651
—111 digits of π

In 1949 the mathematician John von Neumann "expressed an interest in the possibility that the ENIAC might . . . be employed to determine the value of π [Pi]." The resulting experiment, conducted over Labor Day weekend when the machine was not "working" (for the army, to calculate explosions and ballistics), inaugurated the application of electronic computing to the realm of pure mathematics.[1] That weekend π was calculated to over 2,000 digits, which was the furthest computation to date. "To date" in this context refers to thousands of years of manual computation efforts distributed across cultures and continents, a global (if uncoordinated) human knowledge project that has spanned most of recorded history. Then computers arrived and sped things up.

Computational speed increased sevenfold in the twenty-five years between 1939 and 1964, enabling a massive growth spurt, centered in the 1950s, in pure mathematics.[2] The result was a revolution in the field of mathematics for its own sake: five years after von Neumann's experiment, the Watson Scientific Laboratory computed π to 3,093 decimal points. In another five years, in 1959, François Genuys calculated 10,000 decimals of π using an IBM 704 computer in Paris,[3] then in 1962 (drumroll) π was calculated to 100,000 digits by Daniel Shanks and John Wrench using the next state-of-the-art machine: the fully transistorized IBM 7090.[4] IBM's 700 series was the

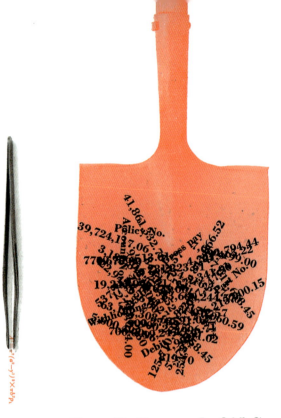

*does both jobs equally well – Burroughs 205 Computer*

*Fig. 7.1:*
BURROUGHS CORPORATION.
PROMOTING THE BURROUGHS
205'S CAPACITY TO HANDLE
BOTH MATHEMATICAL
CALCULATIONS AND DATA
PROCESSING (*MISSILES AND
ROCKETS*, 1959).

first large-scale electronic computer series designed to specialize in mathematical and scientific problem-solving (as opposed to data processing). It had started in 1953 with the 701, a vacuum-tube machine, and over the course of the 1950s the series made the transition from electromechanical to fully electronic computing.[5]

The expansion of the human sense of scale engendered by electronic technologies played out across multiple realms of science and art. The calculation of $\pi$ is one example from mathematics that neatly expresses the advances that automatic, digital computing contributed to a domain with close ties to the computational origins of computer science. It also introduces another graphic symbol system, that of formal mathematics, which was among the raw materials available to artists who sought to depict the new technologies. Mathematical symbols, together with alphabetic language, formed a reservoir of graphic elements (typographic elements, specifically) that artists used in addition to abacuses and dot patterns to convey the rapidly changing identities of computing systems.

Electronic computing and electronic data processing machines may have been in different corners of the industry during the earliest years of their development, but the implementation decade of the 1950s saw a trend toward combination of their functions, as we have seen, into mainframe computers. Artists making commercial artwork were drawn upon to convey the differences between, and the combinations of, mathematical and data processing applications that each new machine performed. The new industry was in rapid flux, and depictions of mathematical and scientific language invoked both its origins and its possible futures. Alongside symbolic and numerical elements, the alphabet also helped to communicate computers' respective strengths.

During this period of identity-building, the range of intersections between computing and language (broadly conceived), when piled, made a vertiginous heap. Typographic artwork was the simple through-line that united many of the language-based phenomena of teenaged computing systems. Computer programming languages flourished and were developed to higher and higher order. Computers began to read, and began to speak; by the mid-1960s some had language-displaying CRT monitors integrated into their design. In response to these prompts, typography became an essential tool for corporate visual communication. The focus of this

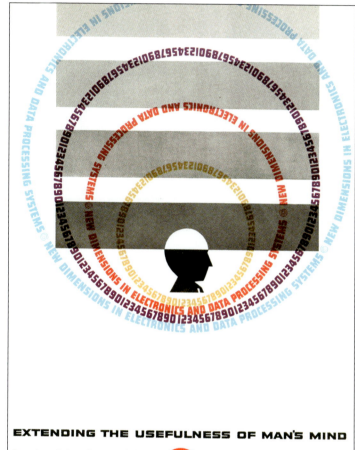

## EXTENDING THE USEFULNESS OF MAN'S MIND

Burroughs continuing, all-out research is producing one newsworthy advance after another in electronics and data processing. And everyone from the owner of the small business to the management of the sprawling corporation is benefiting from them. Here is the equipment that is increasing the speed of data processing and reducing its cost remarkably. That is removing

*Fig. 7.2:*
BURROUGHS CORPORATION. PUBLISHED WHEN THE 205 WAS ITS LEADING PRODUCT, THIS AD MAKES THE SAME POINT ABOUT BURROUGHS MACHINES' MULTIPLE CAPACITIES, BUT FRAMES THEM IN MORE GENERAL TERMS, NOT REFERENCING A PARTICULAR MACHINE. THE ARTWORK IS NOTABLE FOR ITS EVOCATION OF RUSSIAN CONSTRUCTIVIST TYPOGRAPHIC STYLE AND ITS ARTFUL COMBINATION OF ALPHABETIC AND NUMERIC ELEMENTS (*BUSINESS WEEK*, 1959).

COMPUTATION FOR THE SPACE AGE

EXPEDITIONS INTO SPACE FOLLOW TRAILS BLAZED BY COMPUTATION SPECIALISTS. IN THIS HIGHLY
SOPHISTICATED TECHNOLOGY, BURROUGHS CORPORATION'S DEMONSTRATED COMPETENCE RANGES FROM
BASIC RESEARCH THROUGH PRODUCTION TO FIELD SERVICE AS PROVED BY PROJECTS SUCH AS THE AIR
FORCE ATLAS. BURROUGHS CORPORATION IS EQUIPPED BY ABILITY AND ATTITUDE TO FUNCTION AS A
TEAM MEMBER—A CLEARCUT RECOGNITION THAT EVEN IN THE REACHES OF OUTER SPACE, THE SHORTEST
DISTANCE BETWEEN TWO POINTS IS SINGLENESS OF PURPOSE APPLIED TO MUTUAL OBJECTIVES.

**B** Burroughs Corporation
*"NEW DIMENSIONS | in computation for military systems"*

*Fig. 7.3:*
BURROUGHS CORPORATION.
COLLAGE FORMED OF
CIRCUIT SYMBOLS (IN RED),
A SCIENTIFIC PHOTOGRAPH,
AND TYPOGRAPHIC ELEMENTS.
NOTE THE SYMBOL π IN THE
LEFT SIDE OF THE SYMBOL
SOUP (*MISSILES AND ROCKETS*,
1959).

phase in representing electronics is on visible language: numerical and alphabetical typographic elements signifying mathematical processes and data processing, cultivated as design features.

These strategies of representation were better situated within the existing discourses of art and design than were the punch cards or paper tape dots. Graphic design, as much as it is art made for reproduction, is also characterized by the integration of text and image in service of message. Among the modern design movements that offered strategies to American industry in the twentieth century, both Italian Futurism and French Cubism had, during the first decade of the century, freed printed language from its conventions.[6] Artists within both movements deployed letterforms as free agents of design, and in doing so disrupted nineteenth-century ideals of consistency and harmony in page composition. They innovated with typography (as in *fig. 4.15*) in order to experiment with the geometry of line and form. Typeface and type size were combined in new ways as a tool for creating poetic meaning. Russian Constructivists, following Cubism and Futurism by just a decade, used large-format block type as architectural elements in posters,[7] a strategy directly imitated in *fig. 7.2*. The Bauhaus school of the 1930s demanded that new typography be created to serve the future of design and communication, and Herbert Bayer was one of its type designers. These schools and others created an early twentieth-century graphic value system in which typography became a central forum for negotiating emerging links between technology and graphic art.

The great innovation of the commercial artists of this era was to deploy numbers and mathematical symbols as essential yet distinct partners to alphabetic forms. Artists tasked with conveying emerging trends in computing drew on existing traditions but altered them to suit this unique purpose. They introduced numbers, symbols, and letters as related but distinct visual motifs. This strategy, applied by many artists, cultivated new nuances within typographic art. The reader's eye could be drawn into a novel visual field and thereby oriented to the whole field of computing, broadly constituted. Distinctions could be made, graphically, between data and mathematical applica-

tions. As with the appearance of circuit symbols, a specialty graphic design clearly *for technology* emerged independent of other typographic artforms. Not surprisingly, this subsidiary realm of graphic design followed the development, in 1951, of the new language arts tradition of concrete poetry.[8] To thumb through the *Anthology of Concrete Poetry* is to skim hundreds of poems composed of graphically arranged type and letterforms, many of which resemble the artworks in this chapter (and in *figs. 2.10* and *4.15*). Concrete poetry was language written to be seen, rather than read. Its noted works bear a strong visual resemblance to the artworks of this chapter, yet they are composed without number forms.

## A RAPIDLY CHANGING FIELD

*Electronics* magazine remained focused on the broadest possible definition of "electronics" long after computing came to redefine the field. Throughout most of the 1950s, the house record of the industry referred to computing as "electronic data processing," and saw it roughly as a power assist that the world of electronics offered to the preexisting and originally independent fields of data processing and mathematical computation. This was, of course, true for a while, at least in computing laboratories of the

**The RPC-4000 Electronic Computing System can help a company drowning in a sea of figures**

If your company's progress towards new products (and fresh profits) is swamped by a rising tide of figure work ... if your company needs a computing system but has been sitting on the fence waiting for the right one to come along—then you should know more about the Royal Precision RPC-4000. The RPC-4000 is an *advanced*, fully-transistorized computing system offering "medium-scale" capability at a surprising small-scale price. It is equally suitable for engineering or business use. It requires no air conditioning, no site preparation. It plugs into any 110-Volt AC outlet. And, with COMPACT, the new compiler: 1) you achieve machine language compatibility with popular large scale systems... 2) you receive the ultimate in automatic programming techniques and... 3) you eliminate cumbersome conversion routines. Royal Precision RPC-4000's are being delivered now. With it, you get the help of a skilled service force with experience in over 450 computer installations. All good reasons, surely, for writing to Computers, Royal McBee Corporation, Port Chester, N.Y. for more information.

ROYAL McBEE / GENERAL PRECISION
ELECTRONIC DATA PROCESSING SYSTEMS

Fig. 7.5 (above):
AD PROMOTING ROYAL MCBEE'S GENERAL-PURPOSE RPC-4000 (*SCIENTIFIC AMERICAN*, 1961).

Fig. 7.6 (right): ROYAL MCBEE. THE LGP-30 (A LIBRASCOPE GENERAL PURPOSE ELECTRONIC COMPUTER) WAS A BINARY DIGITAL MATHEMATICAL COMPUTER DEVELOPED AT CALTECH WITH A LOGIC SYSTEM INSPIRED BY THE THEORIES OF JOHN VON NEUMANN. IT WAS PROGRAMMED THROUGH AN ICONOCLASTIC "ALGOL-LIKE" LANGUAGE CALLED ACT III THAT WAS WRITTEN SPECIFICALLY FOR IT. DESPITE ITS IDIOSYNCRASIES, IT WAS A POWERFUL COMPUTER FOR ADVANCED MATHEMATICS, INCLUDING SYSTEMS MODELING. IT WAS USED BY MATHEMATICIAN EDWARD LORENZ TO MODEL WEATHER PATTERNS, WORK THAT RESULTED IN LORENZ'S FORMULATION OF CHAOS THEORY.[10] ROYAL MCBEE, LIKE UNDERWOOD, WAS A TYPEWRITER COMPANY THAT DEVELOPED A COMPUTING MACHINERY CAPACITY DURING THE 1950S (*BUSINESS WEEK*, 1961).

1940s. The *Proceedings of the IRE* created a dedicated computing issue in 1953, but it took until 1961 for *Electronics* to recognize the new industry that was beginning to dominate its field with a deep survey. In the survey, the magazine noted that in the five years between 1956 and 1961 the number of computer manufacturers nearly doubled, and that computer sales had increased one hundredfold in the eight years since 1953.[11]

At the outset of the 1950s, the computing world was limited to the big players in information processing and electronics: IBM, Remington Rand (UNIVAC), RCA, General Electric, and Honeywell. During the postwar boom a large number of other companies entered the fray. Business machine companies, transistor manufacturers,

*The LGP-30 Electronic Computer begins breaking up figure-work bottlenecks the very same day it is delivered.*

spinoffs, and startups all sought to compete with industry titans for what they could accomplish during what became a late 1950s' and early 1960s' period of industrial redefinition. With the advent of computing as a new industry, its ur-product was no particular industrial application; rather, it was the capacity of electronic, digital machines to turn data into the new spun cloth of "information."

The Burroughs Adding Machine Company was the country's oldest continuously owned business machine company, but in the mid-1950s it lacked a reputation beyond electromechanical adding machines—desktop devices since the nineteenth century. Burroughs' rapid leap into computer design and manufacturing was equaled by the gains in the industry made by National Cash Register Company (*fig. 7.7*) and Control Data Corporation, among others. But Burroughs invested more than many of its competitors in marketing itself, hence its larger footprint in the historical record of graphic advertising. Advertisements such as *figs. 7.1* and *7.2* helped the company promote its new image in the context of a rapidly changing business environment. The series of Burroughs advertisements in this chapter can be read as a set piece in a company's project to redefine itself as a developer of products that could perform the full range of computing tasks. Its leading products were integrated electronic data processing machines, using electronic circuits to support electromechanical systems: the Datatron 205 computer (promoted in the ad copy that accompanies *figs. 7.1* and *7.2*) was the company's first general-purpose computer, developed following Burroughs' 1956 acquisition of Pasadena-based ElectroData Corp.[12] This acquisition bought Burroughs data processing know-how and drew the company's focal point westward, from its base in Missouri to greater Los Angeles. The advertising artwork in *fig. 7.1* uses typographic elements combined with graphic elements to convey the 205's multiple functionalities: tweezers hold a delicate string of mathematical language, while a shovel signifies the heaps of data the machine can process. With 60 percent of its functionality mathematical and 40 percent data processing,[13] the 205 found its market: it was purchased by banks and insurance companies to manage accounts, but was most prominently used by NASA to control rocket and missile launches.

The question of whether to promote the industry as a science-based technology or as data processing was not value-neutral. The Cold War was a science-based war: the atomic bomb arsenal, the development of an intercontinental ballistic missile delivery system to serve that arsenal, the technological architecture of both ground-based and airborne surveillance systems, all were mathematics-intensive industrial pursuits. Computer companies sought to make themselves indispensable to war production and allied industries, in

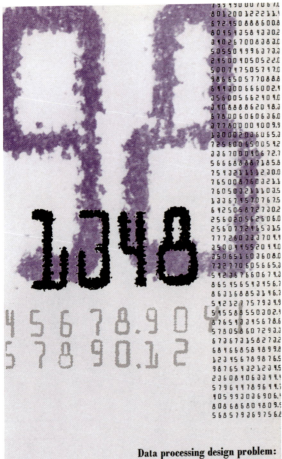

Data processing design problem:

**FOOLPROOF FIGURES**

Thanks to the figures above, designed by the NCR Electronics Division, computer systems can now read "on sight" the printed output of cash registers and business machines. As a result, data processing systems have a key to greater speed, efficiency and economy.

*Fig. 7.7:*
NATIONAL CASH REGISTER
[*SCIENTIFIC AMERICAN*, 1962].

particular the civil space race. Data processing was also important, but the wind in the late 1950s pointed to a future in which science-based capacities and dual science-and-data capacities were essential to the weightiest contracts. Yet, as detailed by Nathan Ensmenger in his history of this era, *The Computer Boys Take Over*, the science-based identity of the emerging mainframe computer industry was not a foregone conclusion.[14] In this context, the preponderance of mathematical and scientific language that dominate the artwork in this chapter reads as clear strategy on the part of computer companies to define their usefulness to wartime opportunities.

**RECRUITING COMPUTER PROGRAMMERS**
There was another reason for promoting the industry in keeping with its laboratory origins. As the industry grew it needed to recruit people with the technical expertise to develop, maintain, and program new computer systems. At the material level the most infinitely small particle inside the machine remained the electron. But at the level of function and signification the new microparticulate was the *code*. Early computers received programming commands via punched cards or paper tape in a coded machine language, or rather, any one of many coded "machine languages." The information theorists Edward E. David and Oliver G. Selfridge expressed their temporary frustration with the irregular progress of human-machine communication in these terms: "As [man] rushes to build his replacements, he notices an interim requirement for man-machine communication. In the meantime at least, computers must be able to, but cannot, understand the writing and talking of [men]."[15]

In a hurry for computers to be able to understand natural speech and read natural script, computer scientists worked for decades toward those goals, which are as yet not perfectly attained. At every step they relied on the efforts of early programmers who wrote and programmed machine code and developed compilers—systems that translate human language into machine language.

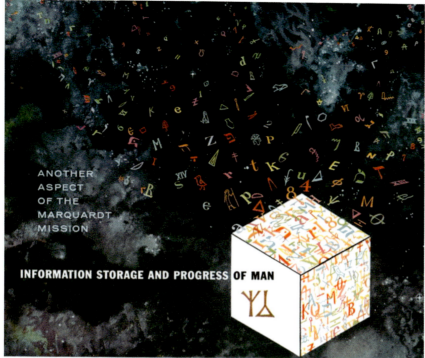

ANOTHER
ASPECT
OF THE
MARQUARDT
MISSION

INFORMATION STORAGE AND PROGRESS OF MAN

If the technological history of man's accomplishments were divided into a twenty-four hour day, the accomplishments of the twenty-fourth hour would vastly outweigh the sum total of the first twenty-three. This surge of progress has resulted in part from man's ability to permanently record and use information relating to his experience.

The writing of human experience — the first permanent information storage system — began only 10,000 years ago. The pages of history written since, would require a storage area larger than the Empire State Building to contain them. Thus man's ability to chronicle facts has far exceeded his capability to store them for easy reference.

Now scientists in the Pomona Division and ASTRO, Marquardt's division for research into the space age, are developing a multi-channel information storage system with an order of magnitude superior to any known storage method. RESULT: all information recorded during the first 100 centuries of civilization's history may be stored in a 6 foot cube.

The future, viewed in the perspective of this new memory potential, offers man the opportunity for an even greater rate of progress — wherein it is possible to envision the achievements of the last minute of the exemplary 24th hour exceeding the sum-total of the first 1,439 minutes.

New information storage concepts typify but one aspect of The Marquardt Mission.

*Creative engineers and scientists are needed.*

ASTRO DIVISION

THE *Marquardt* CORPORATION

CORPORATE OFFICES, VAN NUYS, CALIFORNIA

◆ ASTRO ◆ COOPER DEVELOPMENT DIVISION ◆ OGDEN DIVISION ◆ POMONA DIVISION ◆ POWER SYSTEMS GROUP

$E = IR$     $F = MA$     $B = WT \log_2 \left(1 + \frac{S}{N}\right)$

$\pi r^2$     $\frac{1}{F} = \frac{1}{P} + \frac{1}{Q}$

Early machine languages were differentiated, like the machines themselves, between those oriented to mathematical computation and those oriented to data processing. Programming languages are formed out of algorithms—sets of instructions that cause a computer to perform a particular computing or data processing task to accomplish a specific objective.

$$R_{ij} - \tfrac{1}{2} g_{ij} R = O$$

DATA PROCESSING DIVISION **IBM**.

Languages are varied across a hierarchy of "lower-order" to "higher-order," with lower-order being closer to machine language and higher-order offering a range of programming tools, such as automatic subroutines (preset decision instructions applied to logic gates) and alphabetic coding (instructions given through input/output technologies such as punch cards or paper tape). Compilers were sets of instructions to translate different languages to different particular machines, enabling general-purpose machines such as the Royal McBee machines advertised in *figs. 7.5* and *7.6* to run languages popular at the time, such as ALGOL (ALGOrithmic Logic language) and FORTRAN (FORmula TRANslating language).[16]

Grace Hopper, a programmer of the ENIAC computer who had worked under Eckert and Mauchly, was the first to develop a high-level computer language (one featuring automatic subroutines and alphabetic coding).[17] Her Common Business-Oriented Language, or COBOL, developed in the late 1950s, expanded the usefulness of programmable computers and advanced the capacity of data processing languages and mathematics-oriented languages at the tier of higher-level languages. COBOL was introduced to the commercial market in 1961 by IBM, which packaged COBOL compilers and processors with its 1400 series, the company's early 1960s-era flagship machines for business and industry.[18]

The profession of programming developed along with the proliferation of languages and of working computer systems themselves. In the laboratory years of the 1940s programmers numbered only in the tens, or perhaps dozens, worldwide; a good number of them were women, a legacy of women working as human calculators during wartime. By the late 1950s organizations that wished to compete effectively for technology-based contracts needed to build highly skilled workforces. While as late as the early 1960s there were still many dozens of women working as programmers across the country,[19] the profession was being gradually masculinized. Many print advertisements went beyond promoting a company and its products to become recruitment posters as well, featuring wide-ranging calls for help from "computer people." *Fig. 7.8* is a typical, typographically rich piece of graphic art designed to attract the attention of such specialists. The ad copy accompanying *fig. 7.8* expresses some of the imprecision about areas of specialization within the computer field commonly found in allied industries. Marquardt was an aerospace hardware firm that built propulsion systems for rockets. Its cautious language in this ad, which appeared in *Aviation Week,* asked for "creative engineers and scientists"—likely expressing a general desire for computerization rather than a detailed understanding of the various roles of engineers and programmers.

Some advertisements targeted programmers exclusively,[20] but typically computer programming, being the newest and least well-defined profession among a range of necessary specializations, was mentioned in the context of advertising copy focused on the totality of skills that were needed, as in *fig. 7.9*; the gender inclusivity ("men and women") of this ad was, unfortunately, very unusual. The term "computer programmer" was just emerging at the time. Though it eventually went mainstream, it was more prevalent in early years on the pages of trade periodicals dedicated to the computer industry, such as *Computers and Automation* (founded in 1952 as *The Computing Machinery Field).*[21] In the trade literature of allied industries it took longer for general terms to yield to this more specific new category of labor.

**TEACHING COMPUTERS TO READ AND TO SPEAK**    In 1955 and 1960 the integration of humans and machines in the realm of language took another step. Two, in fact: in 1955 computers began to learn to read, and in 1960 they began to learn to speak. These

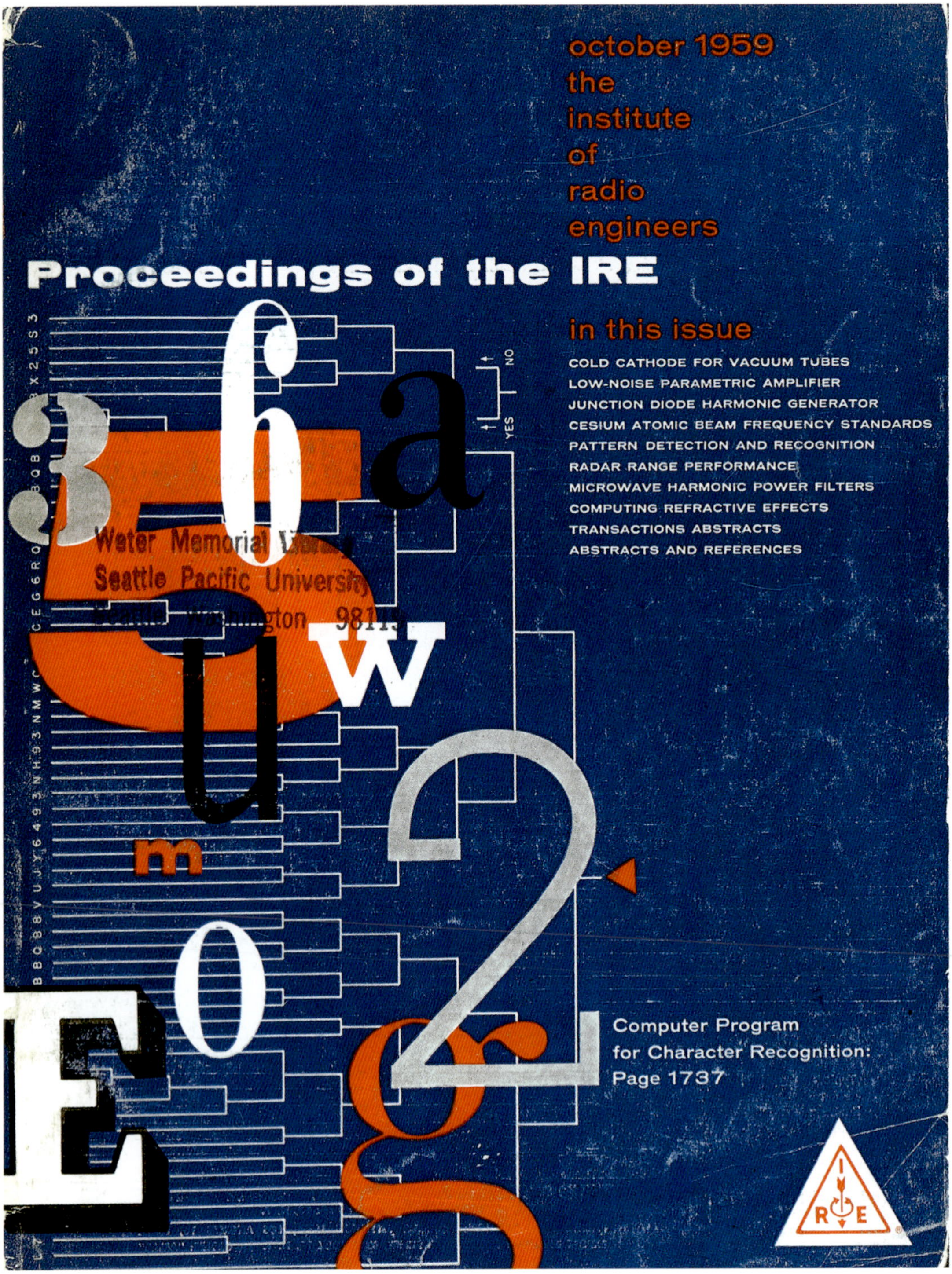

october 1959
the
institute
of
radio
engineers

# Proceedings of the IRE

## in this issue

COLD CATHODE FOR VACUUM TUBES
LOW-NOISE PARAMETRIC AMPLIFIER
JUNCTION DIODE HARMONIC GENERATOR
CESIUM ATOMIC BEAM FREQUENCY STANDARDS
PATTERN DETECTION AND RECOGNITION
RADAR RANGE PERFORMANCE
MICROWAVE HARMONIC POWER FILTERS
COMPUTING REFRACTIVE EFFECTS
TRANSACTIONS ABSTRACTS
ABSTRACTS AND REFERENCES

Water Memorial Library
Seattle Pacific University
Seattle, Washington 98119

Computer Program
for Character Recognition:
Page 1737

advances in computer programming represented incursions into the practical realm from the largely theoretical field of artificial intelligence (AI).[22] While the automated decision trees of early switch systems were technically a rudimentary system of AI, the theory behind what came to be recognized as a field of academic inquiry occurred at the intersection between theoretical mathematics and communication theory. This environment, which flourished in the 1940s (more on it in chapter 9), contributed to the incubation of early philosophies of computing. AI theory left the blackboard in stages, and the integration of cathode-ray tube monitors into computers that began in the late 1950s offered a physical venue for machines to adopt processes that were visibly connected to human processes. CRTs could display machine code and other input/output processes in readable type, and thereby reinforced the notion of computers as machines with language interfaces.

The earliest efforts at electronic automatic "reading," aimed at assisting the blind, occurred as far back as the mid-1930s.[23] In the 1940s, as we saw in chapter 2, a phototube was developed to convert a light-interference pattern (the text on the page) into an electronic vibration that blind people could "read" with their fingertips.

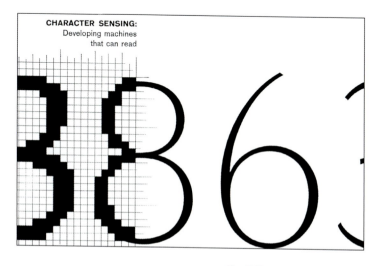

Fig. 7.11:
IBM (*PROC IRE*, 1961).

The first experiments with programming pattern recognition started in the 1950s both in England and in the U.S. at Lincoln Laboratories, IBM, and MIT.[24] (The term "pattern recognition" has its origins here, in computer vision; since the 1960s the term has expanded to include all kinds of pattern-seeking programming including [nonvisual] data analysis.) Some of these experiments were motivated by the use of machines to better understand the human learning process; others were focused on programming computers to replicate the human activity of reading. These early experiments sought to answer the question: Could a computer be programmed to distinguish a mark, or a blob, from white space? In order to accomplish these tasks, machines had to be programmed to superimpose a grid over a perceptible field, and then to evaluate each box within the grid as a binary value, for the presence or absence of a variation. The computer would use the grid results to assemble a binary picture of "there/not there" data points, and

*(opposite) Fig. 7.10:*
PROCEEDINGS OF THE IRE COVER ART, FEATURING A STYLIZATION OF THE DIAGRAM OF THE BINARY LOGIC SYSTEM FOR STEPHEN UNGER'S CHARACTER RECOGNITION PROGRAM (1959).

could then analyze the resulting pattern against a programmed set of shapes corresponding to printed letters.

By 1959 Stephen Unger of Bell Laboratories reported on the state of programming research in the *Proceedings of the IRE* (*fig. 7.10*).[25] His associate, Miss D. M. Habbart, had programmed an IBM 704 tube-driven mainframe to behave as a spatial computer—one able to conduct both pattern detection and pattern recognition. Together, these two programmed operations allowed the logic system of the machine to scan and interpret two-dimensional information—letterforms—that had been previously incomprehensible.

This research introduced the idea, however illusory, of reciprocity between human and machine: humans could "communicate" with machines through programming them; in "reply," machines could mimic the human process of reading. It was a superficial mimicry, as the computer's output was, at best, programmed to reproduce a digital letterform that correctly corresponded to the given input. More significant is the attempted integration of an organic process into the functionality of computing. Theoretical dialogues about AI had a new hook to hold on to, and the proposed integration of human and machine capabilities took a big leap forward. This was the antecedent to today's optical character recognition tools.

*Fig. 7.12:*
BELL LABORATORIES. NOTE THE PAPER TAPE DOT DESIGN, INTEGRATED WITH A MAGNETIC TAPE REEL AND THE UP/DOWN OF A CRT DISPLAY OF OSCILLATING SOUND WAVES. DOMINATING THE IMAGE IS THE LANGUAGE ART CONTRASTING THE PHONEMES PROGRAMMED INTO THE COMPUTER—"H", "EE," "S," "AW," ETC.— WITH THEIR REGULAR APPEARANCE—"HE SAW THE CAT" (45RPM RECORD, 1963).

In the 1969 film *2001: A Space Odyssey* director Stanley Kubrick offered up philosophical dialogues about AI in the format of the most expansive quasi-experimental film that 70mm cinema has ever seen. In the space epic, the computer HAL (short for Heuristic Algorithmic computer) sings the song "Bicycle Built for Two" as it is being dismantled. HAL's plaintive attempt to express itself with "humanity" coincides with the revelation that it also expresses a very human type of *inhumanity*: it is being punished for having murdered a crewman. One of the film's central narratives is a simple provocation: that we should be very careful what we wish for when we begin to program computers to behave like human beings.

The recording of "Bicycle Built for Two" heard in the film is the same one that is on the 1963 Bell Laboratories record *Computer Speech* (*fig. 7.12*). The record was published by the lab for use in classrooms, though it entered the culture as a novelty item and was widely circulated. Research into synthesized speech had previously been based on analog speech synthesizers; Bell Laboratories' innovation was to simulate synthesized speech using a "high speed, general purpose computer"—actually IBM's first all-transistor general-purpose computer, the 704—that had been programmed with punched cards to make sounds corresponding to twenty-two consonant tones and twelve vowel sounds.[26] This research was framed as part of a larger project to build long-distance telecommunications tools, so that people might someday be able to type on a keyboard and have a computer thousands of miles away enunciate what was spoken. It was a startling incursion by a machine into the human-machine interface zone of language—a startle factor that Kubrick relied upon to build suspense in *2001*.

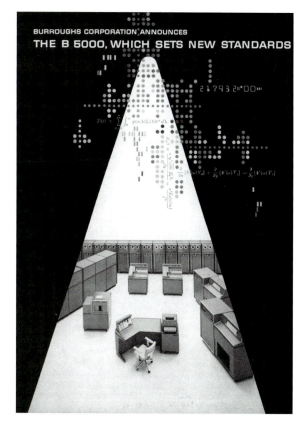

## THE BURROUGHS B 5000

The Burroughs B 5000, released in 1963, was a mainframe computer that nudged the industry ahead a few steps with architecture and design innovations. It was the first computer to have its logic system wholly simulated on another machine (the 205) before it was built. In other words, it was virtually field-tested. It was also the first large multipurpose computer to have its own encoded ALGOL compiler, the first bundling of software—computer language—with hardware.[27] Prior to the B 5000, hardware was the only permanent fixture of a new computer. The B 5000 was developed between the late 1950s and 1963, though heavy publicity around the machine stirred in the summer of 1961. The machine was a gamble for Burroughs, which was trying to catch up to IBM and UNIVAC in the mainframe computer business.

In the 1961 magazine advertisement (*fig. 7.13*), mathematical formula strings are interwoven horizontally and vertically with 7-bit paper tape dot patterns, punched card dots, and numeric strings associated with banking industry data processes.[28] These elements are gracefully arranged in a

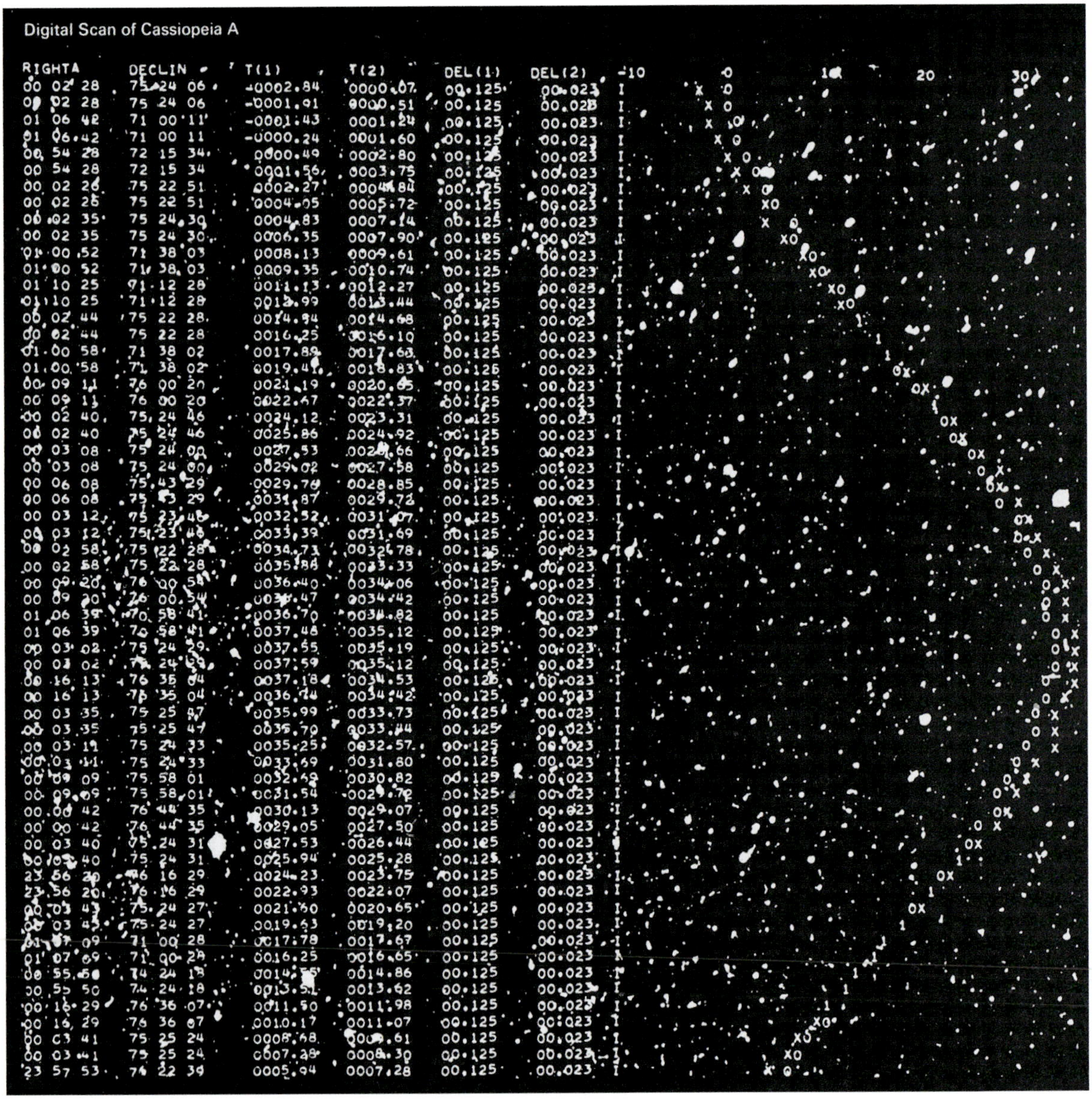

Fig. 7.14:
JACQUELINE CASEY FOR THE LINCOLN LABORATORY. NUMERIC CODE REPRESENTING A RADAR ASTRONOMY DIGITAL SCAN OF THE CONSTELLATION CASSIOPEIA IS LAYERED AGAINST A TELESCOPIC PHOTOGRAPH OF THE CONSTELLATION AND ITS VICINITY (*PROC IRE*, 1966).

lacy curtain that drapes in front of the unglamorous mainframe array. The image both encloses and reveals the machine; it is dominated by the data-and-language curtain that demands our attention, yet the bird's-eye panorama of the large machine is also striking. The parted black drape mediates between the "front" (language curtain) and "back" (machine photograph) of the image. It's a graphic visualization of the tension between what's *inside the machine*—language of different kinds, nothing less than the machine's entire use value—and its gawky exterior. The result is a form of industrial theater, bringing to mind the elaborate sets and inventions of both stage and film.[29]

This association relates to the tension between graphic design and the insistent emergence in the 1960s of wide-angle photography, a medium that would soon challenge other art forms for dominance in commercial artwork. The artwork also reflects the contrast between the giant mainframes of the early 1960s and the simultaneous advances in circuit miniaturization. Many devices did get smaller because of miniaturization. The navigation and communication tools of avionics and space electronics were vastly aided by miniaturization, while they also became parts of larger systems, tethered by a radio or data link to a mainframe control computer on the ground. The narrative of miniaturization tended to drive public understanding of developments in electronics as it affected the look and feel of everybody's new radio. As a result the new mainframe computers, which did not face out to the public anyway, were a struggle for artists to interpret. The black "curtains" of the B 5000 advertisement both part to reveal the new machine and enclose it, letting the visible language and paper tape dots convey the ad's message.

In the previous chapter we saw that the combination of electromechanical data processing technology with electronic circuits and with binary mathematical calculation together yielded the "age of information." This observation reflects a material change in the language around computing that played out in many contexts as the new industry developed and branded itself.[30] The case of the B 5000 offers an additional view to the emergence of this thinly constructed "age": In an article about the new machine in *Data Processing*, the scientific processor within the machine is referred to as a "mathematical" processor, while the machine's dual capacity in natural language data processing is described independently. Only the whole system, integrating both capacities, is described in the article, according to Burroughs' corporate literature, as an *information* processing system.[31] In this view, the transition from data to information takes place when alphabetic and numeric modes of computation are brought together within a common machine rubric.

*Figs. 7.14* and *7.15* were created in 1963 by Jacqueline Casey (1927–91), an eminent female graphic artist of the twentieth century. It is serendipitous for this book that Casey was responsible for creating artwork for recruitment advertisements for Lincoln Laboratory. Founded in 1951 in response to postwar developments in electronics, the laboratory was organized by MIT at the request of the military and its Advanced Research Projects Agency (now DARPA). The lab's Cold War purpose was to harness the academic research environment of MIT to the task of integrating state-of-the-art science-based electronics research with U.S. military technologies.[32] In practice, the lab became a center for the development of radar astronomy, pursuing pure astronomical science on the side in addition to developing Cold War radar-based surveillance installations.[33] The SAGE System (Semi-Automatic Ground Environment System) and the DEW Line (Distant Early Warning Line) were its first two projects, both major architectural elements of the Cold War.

Casey, a native of the Boston area and a graduate of the Massachusetts College of Art and Design, started as a graphic designer at MIT in 1955. She had the good fortune to have Muriel Cooper as a colleague at MIT's Design Services Office (then the Office of Publications). Cooper would go on to become head of design at MIT Press, and would later found the Visible Language Workshop at MIT's Media Laboratory, pioneering the study of computer graphics and communication.[34] Casey also had a significant impact on the relationship between electronics and graphic design, though her work is less well known than Cooper's. Casey was taught at MIT by the visiting Thérèse Moll, a Swiss graphic artist who introduced Casey to the major principles of Swiss graphic design, including its focus on the integration of typography and graphic art.[35]

Casey created MIT's posters and graphic advertisements announcing exhibits, lectures, and other events for thirty years.[36] Among her additional, lesser-known works are at least twenty-five graphic recruitment advertisements for Lincoln Laboratory (including *fig. 4.10*, in addition to those in this chapter). Dominated by solid black backgrounds that bleed to the edge of the page, the works in Casey's Lincoln Laboratory series all use visual elements that originated with lab scientists, each representing some aspect of electronics research. Most, such as *fig. 4.10*, are stylized micro- or macro-photographs. Casey was tasked with developing recruitment advertisements that conveyed the following to potential recruits:

- the mission and research areas of the laboratory
- its affiliation with MIT
- characteristics desired of applicants, and the urgency of the need for new staff
- "intangibles, such as prestige, 'atmosphere'"[37]

*Fig. 7.15*, however, shows the influence of the Swiss design tradition on Casey's work. In it, she uses typographic artifacts as contrasting elements within a larger design, and organizes that design in a grid structure, a major principle of the Swiss tradition. Eight distinct vertical columns of text and graphics from at least nine different scientific papers are combined,[38] representing current laboratory research in radio astronomy, solid state physics, and circuit design, as well as astronomical physics. A block diagram of a computer logic system (left edge) draws the eye of computer programmers, as does the mention of the IBM 709 in the adjacent column. Casey has arranged the texts, the scientific diagrams, and the mathematical equations from the cluster of papers into a total design, greater than the sum of its parts.

The design's unique aesthetic explores three distinct points of tension between art and electronics. First is the combination of solar system physics and solid-state electron physics, represented in the technical language of some of the papers. The close juxtaposition of these two domains of inquiry forms a very different kind of exploration of scale than Herbert Bayer's "Earth bulb," yet it derives from similar underlying motivations. Second, contrasting columns of text draw the eye through adjacent serif and sanserif blocks of text. In this placement, Casey is playing with the defining modernist typographic movement away from decoration and toward sanserif type. Her design creates a framework where the two type styles complement each other and suggests a reciprocal relationship between the past and the future in type design, and perhaps by extension in science as well.

Casey's strategy of playing with "found" texts comments on the limitations of design to influence the world in which it operates. Her collage style introduces the third point of tension, which is between the viewer's perception of scientific and technical literature—that perhaps it is not art—and the offered evidence that it *is* its own art, and does not need "art" to be made about it. Casey has chosen physics papers that include diagrams of lunar contours and combined them with diagrams of magnetic symmetry patterns within electrons. The soft center to the image introduced by these two curvilinear diagrams suggests that the juxtaposition of physical inquiry with

le form o The general spin v
a simultaneous sol
the magnetization
In contrast to prev $- \zeta^*(t_1)] [\zeta(t_2) -$
conductivity, rela
are all properly ta
algebraic equatior
of the wave numbe $_o(t_1 - t_2) + \dfrac{R_{10}}{2}$
lutions. Some of i
ing plane waves w
ones in the directi ie $\rho(t_1 - t_2)$ of th
analytical solutior
the special case w
direction of the st
solution for the ca
ient obtained on $_1(t_1 - t_2) - \dfrac{[1 +}{}$
lutions for the latt $_o(t_1 - t_2)$
using an IBM 709
ative results are gi
When relaxation a
lected, our result f likelihood ratio
Kittel in the static some algebra,
found that the unil
truly exist only un
namely, under the
one of the two nor $\zeta(t)\,\mu_2^*(t)\,dt - -$
dium. $\qquad 1$

"Excitation and Bc2
Resonance"

J. Appl. Phys. 32, $\dfrac{2}{\lambda} \displaystyle\int_o^T\!\!\int_0^{\,} \zeta(t_1)\,\mu_1^*(1$

Using the general
the general spin w
tion of the line sh is indicated in F
metallic films is m
conditions at the s l parameters knc
set of excitation c
lation show that, \
spacings between t $(i) = \mu^\bullet(-1,\bar{\gamma}_2)e^{-i\omega_o t}$
peaks are determir
stant A, the relati
onances are strong
sotropy energy der
Integratc.

coupling resu.
along the [001
"extra" $\gamma_5$ ele
tortions would
directed along
partially occu
$(\alpha < 60°)$ dist
$(c/a < 1)$ CoO

because of hig
and the symm
been prepared
random throug
Although these
$T_N$, they prob

Since $\Delta_{LS}$
distortions in
0.91),[78] and C
in FeO and Co
Nevertheless,
$Fe^{3+}, Ni^{2+}, [Ni$

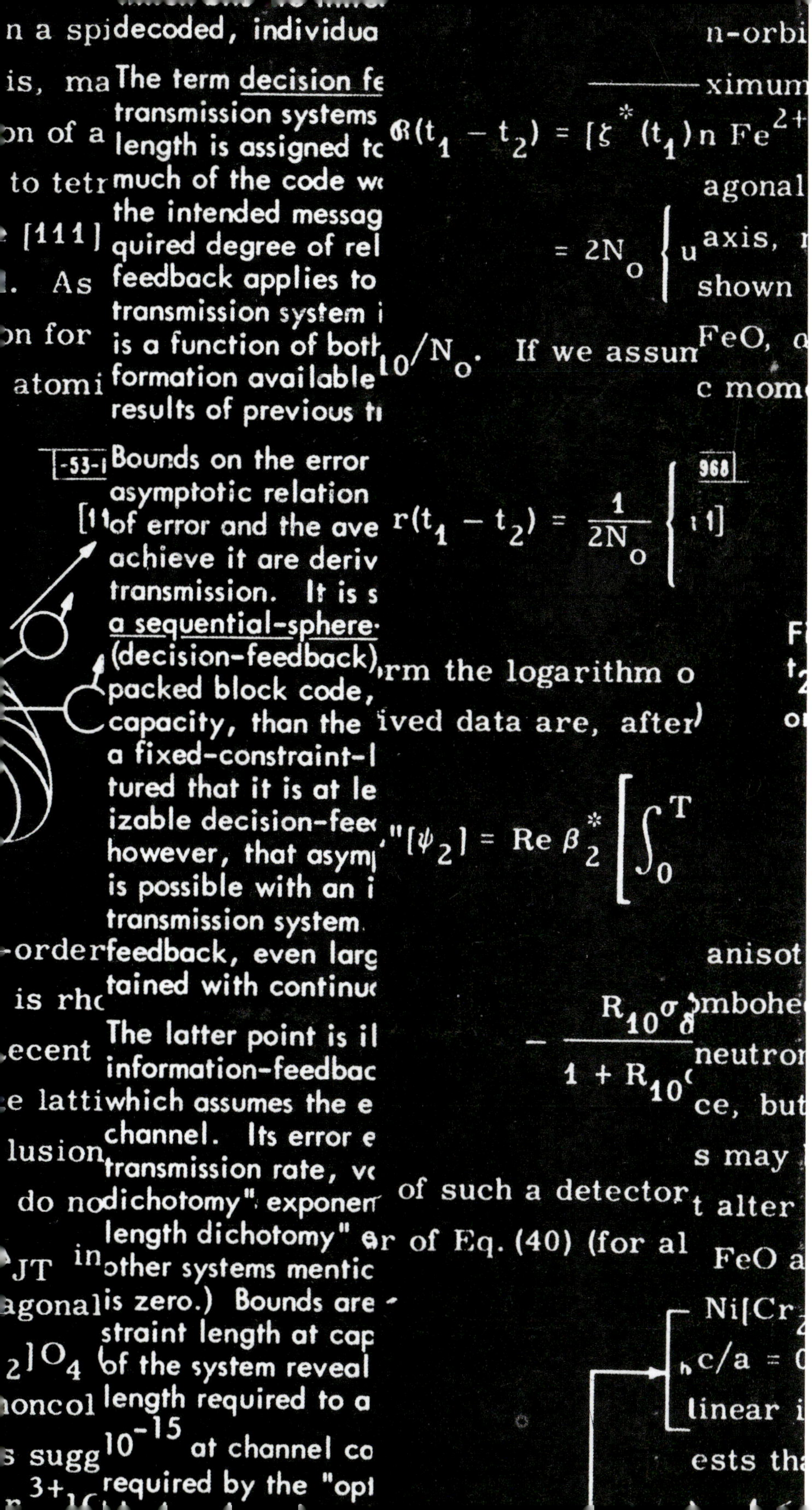

n a spi decoded, individua                                        n-orbi

is, ma The term decision fe                          ─────── ximum
       transmission systems
on of a length is assigned tc   $\mathcal{R}(t_1 - t_2) = [\xi^*(t_1)\,n\,Fe^{2+}$
to tetr much of the code w
       the intended messag
[111]  quired degree of rel             $= 2N_o \begin{cases} u \text{ axis,} \\ \text{shown} \end{cases}$
.  As  feedback applies to
       transmission system i
on for is a function of both $_{tO}/N_o$.  If we assun FeO, a
atomi  formation available                              c mom
       results of previous ti

[-53-i] Bounds on the error                             [968]
        asymptotic relation
[1] of error and the ave  $r(t_1 - t_2) = \dfrac{1}{2N_o} \begin{cases} i\, i \end{cases}$
    achieve it are deriv
    transmission.  It is s
    a sequential-sphere·
    (decision-feedback) rm the logarithm o                F
    packed block code,    ived data are, after)           $t$
    capacity, than the                                    o
    a fixed-constraint-l
    tured that it is at le
    izable decision-fee   $"[\psi_2] = Re\,\beta_2^* \left[ \int_0^T \right.$
    however, that asymi
    is possible with an i
    transmission system.
-order feedback, even larg              anisot
    tained with continuc    $R_{10}\sigma$ ombohe
is rhc                                  neutror
ecent The latter point is il   $-\dfrac{}{1 + R_{10}c}$
       information-feedbac
e latti which assumes the e                ce, but
lusion channel.  Its error e
       transmission rate, vc              s may
do no  dichotomy", exponer  of such a detector t alter
       length dichotomy" er of Eq. (40) (for al FeO a
JT in  other systems mentic
agonal is zero.)  Bounds are    Ni[Cr
        straint length at cap
$_2]O_4$ of the system reveal      $_hc/a = c$
noncol length required to a
s sugg $10^{-15}$ at channel co        linear i
3+     required by the "opl           ests tha

Fig. 7.15:
JACQUELINE CASEY FOR THE
LINCOLN LABORATORY (PROC
IRE, 1963).

# You can explore new areas at IBM in

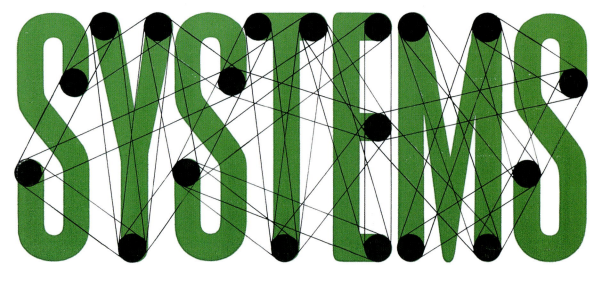

the surrounding linear equations encompasses the range of geometric and organic approaches to understanding the universe.

**COMPUTER SYSTEMS**    Historian Paul Ceruzzi has pointed out that "electronics" as a broadly defined field was terminally redefined when computers took over the field in the 1950s: "electronics" came to *mean* computer science, while noncomputing electronics assumed secondary roles as affiliated technologies.[39] The development of microprocessors from circuit boards and the development of integrated software/hardware computers from adding machines are among the material advances that pushed this transition. Systems engineering that connects computer-driven controls systems to other advanced electronics projects is another such gesture. Lincoln Laboratory's SAGE System was powered by an IBM computer system. Both the SAGE System and the DEW Line integrated advanced electronics such as radar systems and CRT monitors with computer systems that coordinated their work across thousands of miles. The Cold War was a war of surveillance, and as its technological systems matured they demanded automated processes for transmitting and interpreting information.[40]

The Lincoln Laboratory was therefore a site of innovation in computing: magnetic core memory was developed there, and the laboratory's work demanded computing systems that could process inputs from dozens of radar arrays simultaneously. IBM's capacity to develop the necessary systems firmed the company's anchor position as a purveyor of Cold War electronic technologies, a position that advanced its status still further beyond that of its competitors. The integration of computing systems with other advanced electronics devices in service of the Cold War made the Lincoln Laboratory an emblem of 1960s' computing: a developer of systems that had crossover applications in civil space exploration.

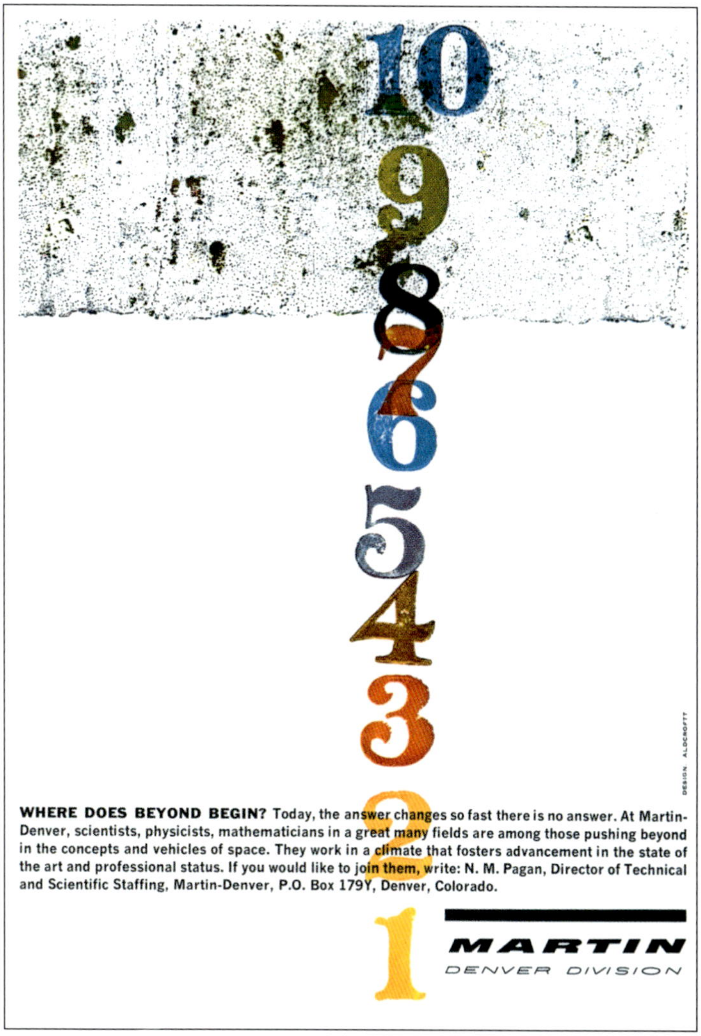

**AUDIBLE LANGUAGE, VISIBLE LANGUAGE**

In the spring of 1969, the English poet Neil Mills wrote the first in a series of number poems, sequences of numbers written to be read aloud for poetic effect. Mills noted, with regard to his development of number poetry, that numbers were "limited" in their poetic potential; however, "if organised into certain juxtapositions and rhythmic breaks, and read with the regard for pitch, volume and sensitivity accorded to more traditional poetry reading, they could be made to yield an unexpected lyrical or evocative content."[41] Mills was writing only about his own work, not about the countdown to a rocket launch. Forty-five years after the fact, however, it is impossible to read his remarks outside the historical context of that particular moment in the relationship between arts and sciences. His effort to cultivate this new genre followed two decades in which mathematics, among other disciplines, was expanded literally to new dimensions through its engagement with electronic computing.

In addition to the excitement around the ever-expanding calculation of $\pi$, computing made possible advances in cryptographic science, such as the

WHERE DOES BEYOND BEGIN? Today, the answer changes so fast there is no answer. At Martin-Denver, scientists, physicists, mathematicians in a great many fields are among those pushing beyond in the concepts and vehicles of space. They work in a climate that fosters advancement in the state of the art and professional status. If you would like to join them, write: N. M. Pagan, Director of Technical and Scientific Staffing, Martin-Denver, P.O. Box 179Y, Denver, Colorado.

MARTIN
DENVER DIVISION

*Fig. 7.17:*
THE MARTIN COMPANY. THE ORIGINAL NUMBER POEM: COUNTDOWN TO LAUNCH (*MISSILES AND ROCKETS*, 1960).

otherwise unattainable string represented by the Rand Corporation's Million Random Digits project. Such projects were a bit obscure, yet the Million Random Digits were published in book form by the Free Press in 1955, and as such formed a numerical literary work that was arguably better known than the era's adjacent schools of poetry.[42] The umbrella term within which Mills and his fellow poets cast their collective works was "Experiments in Disintegrating Language," a phrase that gains resonance with a historical gaze that aligns this poetic "experiment" chronologically with the introduction of scientific, technical, and programming languages into a broader context and sphere of significance than they had inhabited prior to 1950.

More particularly, Mills's 1969 work coincided with the climactic year of the three human space-flight programs of the 1960s. Those programs—

Mercury, Gemini, and Apollo—were the capstones to two decades of emergence pertaining to numbers and their arts and sciences. The three programs also brought to the general public through mass media—including international mass media—the regular recitation of both launch countdowns and other alphanumeric strings of information that engineers spoke aloud to one another as a form of program-specific communication.

The abundance of incidental number strings in public media of the 1960s prefigures the intense saturation of incidental alphanumeric texts we live with today. The contemporary poet Kenneth Goldsmith has pointed out recently that with so much language being generated all around us, there is little call for anyone—any poet, at least—to compose new material.[43] His call for a new poetics of navigation, the navigation of textual abundance, calls specifically for people to engage "invisible" language, such as computer code and incidental communications, as raw material for new work. Goldsmith's contemporary conversion of traffic reports into poetry[44] postfigures the association between the number poets of 1969 and the launch countdowns of Project Apollo flights. Goldsmith's work goes further to offer context for Jacqueline Casey's language-based graphic collage of 1963 than the studies of such work that were contemporary to Casey. Goldsmith also points out that computers today burst with encoded language, though their code remains hidden. A computer monitor displaying lines of unresolved code, as in *fig. 7.16*, would today not be a promotional advertisement but a transgression, a rupture in the layers of "skin" that have been carefully wrapped around what's inside.[45]

Typographic art for industry now appears, from our twenty-first century perspective, as an artistic response to the unwrapped code—a response dedicated to exposing language in order to build, rather than unbuild, the identity of new machines.

## CHAPTER EIGHT

# THE FURTHEST HORIZON:
# SPACE ELECTRONICS

Space conquest, intercontinental ballistic missiles——neither of these new technological advances would be possible without a multitude of instruments that extend man's senses; that observe and remember, and compute faster and more efficiently than the human brain under similar circumstances.
                    —Simon Ramo, 1958[1]

**CHANGING INDUSTRIES, CHANGING CONTEXTS**

The arrival of electronics in outer space symbolically resolved the dramatic tension formed when Niels Bohr and Ernest Rutherford originally invoked an alignment between the solar system and the electron in the "planetary" model of the atom. The physicists' investigations had followed Einstein's revelations about the nature of the space-time continuum by only a few years. Rutherford's invocation of the solar system in his descriptions of the subatomic world mapped the electron's structure into a frame of reference bounded on one side by Einstein's investigations into space and time, and on another side, by the future of atomic science.

The artwork that promoted and recruited for space programs offers a chance to consider some functions performed by commercial art in the mid-century that differ from those seen so far. "The space program" is a collective noun that refers here, as elsewhere, to what were several very different programs: the numerous robotic programs, both scientific and military, and the three human spaceflight programs of the 1960s: Projects Mercury, Gemini, and Apollo. All of the 1950s-era programs were, to differing extents, adaptations of the Cold War military space programs that preceded the

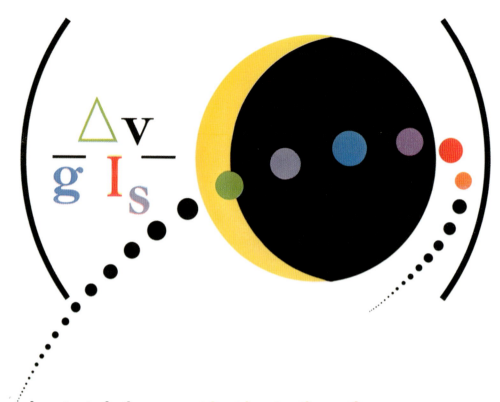

## how to circle the moon without leaving the earth

*Fig. 8.1:*
VISIBLE LANGUAGE: IBM
PROMOTES ITS RECENT
DELIVERY OF TWO 7090
COMPUTERS TO NASA'S
MARSHALL SPACE FLIGHT
CENTER IN HUNTSVILLE, AL.
THE COMPUTERS WERE USED
TO SIMULATE THE ORBITAL
AND LUNAR TRAJECTORIES
OF THE SATURN ROCKET.
THE AD COPY POINTS OUT
THAT COMPUTERIZED FLIGHT
SIMULATIONS WOULD REDUCE
THE NUMBER OF NECESSARY
TEST LAUNCHES BY A FACTOR
OF ONE HUNDRED RELATIVE TO
WORLD WAR II–ERA ROCKET
TECHNOLOGY. THE "HERO"
IN THIS PIECE OF ART IS THE
ORBIT ITSELF (*SCIENTIFIC
AMERICAN*, 1961).

establishment of NASA in 1958. Within these were significant programmatic differences between orbital and planetary spacecraft, and differences in the cultural signification between science-driven programs, classified military programs, and those programs whose central purpose was a cultural one: human exploration. In addition, the launch-less pure science of radio astronomy figures quietly at the center of electronics-based space exploration (see *figs. 2.4, 7.14,* and *7.15*).

As complex as this bundle of programs was, they unfolded nearly simultaneously in the span of a few years, and they drew on a common pool of industry contractors and subcontractors, as well as a common labor pool. The space program gained its mandate and its funding so abruptly—following the USSR's launch of *Sputnik* in 1957—that it had to define itself within the larger spectrum of postwar American industry in a timespan of a handful of years. Project Mercury, in 1961, had forty-seven primary contractors and subcontractors, assisted by an estimated 4,000 suppliers,[2] while in March 1965 the *Washington Star* reported that 500 contractors would work on Project Gemini.[3] The article calls out the role of computer manufacturers, noting that

Burroughs, Honeywell, RCA, and IBM are all among the lead contractors. During those years a half million people were hired into jobs that supported the civil space program alone.[4] (Numbers on military space programs are, even today, less available.)

A full technical account of space electronics is not possible here. This chapter continues to follow the art, and the question: What cultural history of space electronics can be extrapolated from a close look at the associated commercial graphic art? The result is an interplay between technological developments that drew intense public attention and expanded forms of corporate expression. The graphic literature created to promote space electronics altered the dynamic between art and technology, compelling it to engage in cultural macro-narratives of heroism and exploration.

As the space program unfolded, the orbital and interplanetary exploration goals of both the U.S. and the Soviet Union assumed enormous symbolic significance within both countries. Each initially developed its satellite program as a contribution to the International Geophysical Year (IGY)—an eighteen-month window of global scientific cooperation for the purpose of exploring planetary, atmospheric, and space phenomena. During the IGY the missile buildup and espionage competition that characterized the Cold War was abruptly extended to these scientific and cultural realms of signification, and was further amplified by public enthusiasm. The space programs of both countries became nationalist expressions of scientific and engineering one-upmanship, with a noisy subtext of military competition. This atmosphere fed the mood of corporate expression that dominates space-oriented advertising. The stakes were much higher than they had been for other applications of new technologies, and the amplitude of the associated iconography became louder as well.

The rapid redefinition of electronics and computing fields resulted in expanded duties for the business-to-business communications we have seen at work so far. Companies were no longer merely trying to explain their products and sell them to one another. There were additional mandates to fulfill, chief among them the development of a new identity for space electronics as a subset of the electronics industry. Close behind was the need to recruit people who were both space-oriented and electronics specialists. This was a subproject within the larger assignment of space industry recruitment, which cast a wide net across the whole range of American industry and relied on an identity-building display of spectacular visual vernaculars.[5] The graphic literature in this chapter is specific to electronic engineering; as such, the advertising artwork had to balance a focus on space with an emphasis on the urgent

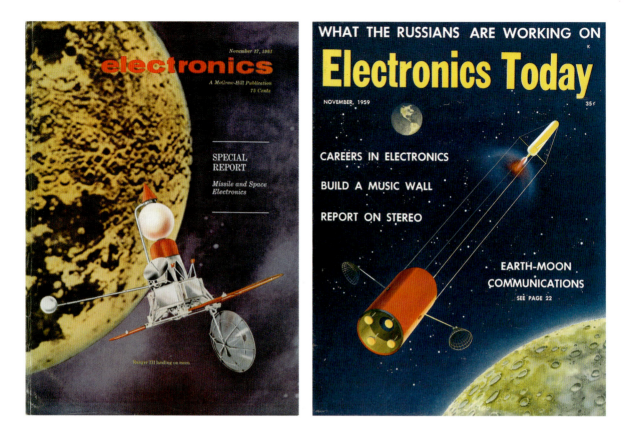

and essential role of electronics. The result is a moderated engagement with science fiction-themed graphic art.

By "science-fiction themed" I'm referring to the subject matter (fantastical spacecraft, as in *fig. 8.3*) and to the process by which graphic art at times created fictional images of proposed scenarios and technologies that in many cases never would come to exist. This fantastical art was the result of an economic gold rush activated by the funding of the space program, combined with the fact that technology was moving so fast at the time that depictions of emerging technologies were driven to outpace their real-world development. The appearance in late 1959 of *Electronics Today* (*fig. 8.3*) exemplifies these simultaneous trends. A new periodical in a crowded field is an artifact of a gold-rush mentality; it positioned itself as a trade magazine but in fact it was a pulpy knockoff. The premiere issue features the cover article as promised, but is written by an author whose name appears nowhere in any technical or related literature. The cover art depicts a "proposed space gondola" that has no attribution beyond the author's imagination. In contrast, the special

"Missile and Space Electronics" issue of *Electronics* magazine in 1961 (*fig. 8.2*) features a rendering of the *Ranger 3* spacecraft and a story about the mission that is packed with technical detail and includes a complete rundown of NASA programs and appropriations. Unfortunately *Ranger 3* outflew its trajectory and overshot the moon,[6] leading to calls for improved navigation electronics.

**GETTING OFF THE GROUND: VACUUM TUBES AND SATELLITES**

Speaking of technical detail, we should draw a line between the first vacuum tubes in space and the space programs of the 1950s. Electronics have been assisting with high-altitude exploration since 1935, when vacuum tubes first entered the stratosphere aboard a craft called *Explorer II*, a stratospheric balloon. Carrying twenty tube-driven "cosmic ray telescopes"—Geiger counters—the balloon was set up, along with a complement of atmospheric sensing tools and radios, to measure electromagnetism and other atmospheric characteristics.[7] The U.S. would enter the Cold War–era space race twenty-two years later with the 1958 launch of the IGY satellite also called *Explorer*, following an interval filled with developments in avionics.

The results returned by the balloon *Explorer II* were only preliminary,[8] but its launch inaugurated the era of electronic investigation into atmospheric

*The role of graphic art.*

*Fig. 8.4 (above):*
THE *VANGUARD* SATELLITE DEPICTED ON THE COVER OF THE ALLIED RADIO CATALOG (1957).

*Fig. 8.5 (below):*
BELL TELEPHONE LABORATORIES. THE *VANGUARD* SATELLITE AS IT ACTUALLY APPEARED (ROUGHLY), WITH LARGE ANTENNAS AND BLOCKY SOLAR PANELS ATTACHED TO AN OPAQUE EXTERIOR (*BUSINESS WEEK*, 1958). BOTH IMAGES PROMOTE THE THREE-POUND SATELLITE'S RELIANCE ON MINIATURIZED PARTS, SPECIFICALLY THE TRANSISTOR.

phenomena. Telegraph signals and radio signals had been subject to interference from electromagnetic radiation in the upper atmosphere from the outset, so an attempt at electronics-assisted research into the sky was an exciting benchmark for both science and the burgeoning electronics industry of the 1930s.[9]

To aviation enthusiasts, the space age had dawned a bit earlier than it did for the general public: experimental aircraft had been pushing the boundaries of sound and space since the late 1940s. In 1959 the U.S.'s best-performing spacecraft was the X-15 airplane, which was actually a spaceplane that had more in common with the Shuttle program that would follow in the 1980s than it did with its contemporaries, the robotic probes and human spaceflight capsules of the 1960s.[10] While this book has tracked the emergence of component electronic parts, aircraft communications and navigations systems spurred the development of some of the most sophisticated and far-reaching devices and systems that those components enabled.

The dozen years that elapsed between the end of World War II and the commencement of the space race proper were a period of intense experimentation in aircraft design, both hardware and avionics. Bell Aircraft had built the first jet in 1942 (with engines by General Electric); five years later its X-1 experimental aircraft broke the sound barrier, and its X-series aircraft went on to set speed and altitude records with successive generations.[11] Bell would then build the navigation system for North American Aviation's X-15 spaceplane in 1959.

When the IGY satellites inaugurated the space age as commonly understood, aircraft companies quickly redefined themselves as part of the new aerospace industry. Lockheed, the Martin Company, Douglas, Boeing, North American Aviation, and Hughes—all companies whose advertisements we've seen in this book—are among the largest aircraft firms that in the Cold War adapted their avionics and hardware capabilities to missile and rocket applications. The aerospace industry was formed in the 1950s from the combination of preexisting aviation, electronics, and computing industries in much the same way that the computer industry was itself formed around the same time from a similar combination of preexisting elements.

For most of these companies, like Bell, it was an easy transition. The constraint of small craft size and the wide range of environmental conditions characteristic of flights by advanced aircraft worked as a rehearsal zone for space environments. By 1958 Bell was already building missiles, and was contracted to build the Dyna-Soar spaceplane (a program that was later canceled).[12] For the satellite age, Bell quickly adapted its most advanced aircraft motor technology to rocketry, yielding the Agena rocket. In 1960 it advertised its contract to power the Air Force's Corona spy satellite series (*fig. 8.6*). Bell also claimed credit for coining the term "avionics." Recall that, in the 1930s, pilots relied on Morse code because their shortwave flight radios were too weak to transmit voice communications. Consider the technological distance traveled between that point in time and the space-age communications and inertial guidance (flight trajectory guidance) systems of the 1950s, and the reason for the emergence of "avionics" is clear. Images such as *fig. 8.6* express the studied revision of identity that many companies undertook when adapting their technologies to space applications.

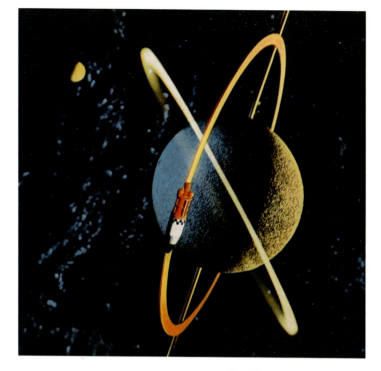

Fig. 8.6:
X MARKS THE SPOT. ENTERING THE SPACE AGE WITH THIS AD PROMOTING ITS AGENA ROCKET, BELL AIRCRAFT SUBTLY PROMOTES ITS HISTORY WITH THE X-SERIES EXPERIMENTAL AIRCRAFT (*MISSILES AND ROCKETS*, 1960).

When Fairchild Semiconductor won the contract to transistorize the B-52 bomber in 1957,[13] that plane probably carried transistors higher than they had gone to date. The orbital space race that the U.S. joined in 1958 was transistorized from the start—on the U.S. side, anyway—and the heavy promotions featuring the *Vanguard* satellite were often stylized to highlight the achievements in electronics miniaturization that it represented (*figs. 8.4* and *8.5*). However, an equal amount of the electronics sophistication that enabled spaceflight was located in ground-based instrumentation. *Vanguard*, for example, was tracked by radar "so accurate that if it was used in a ball game, it could call a six-inch ball hit out of Yankee Stadium 'fair or foul' from . . . 88 miles away," commented the otherwise dry journal *Electronics Engineering* in its report of the satellite's launch.[14]

**COMPUTER SYSTEMS** As we've seen, the dual forces of miniaturization and large-format computing worked alongside each other to change the shape of electronics components during the 1950s. Transistorization and circuit board miniaturization made many devices smaller, lighter, and more flexible than could have been imagined at the close of World War II, while the new computing machines grew to fill entire rooms. These developments were essential to the respective space programs conducted by the army, the navy, and later NASA. Each program would need both the large and the small in abundance.

In fact, none of the space-based applications of electronics would have stayed aloft without the systems developed by the emergent computer industry over the course of the 1950s. At the practical level, space applications called for the most advanced systems engineering that the computer industry had yet come up with. IBM and Burroughs, two heavyweights of integrated information processing systems, both gained contracts for the development of space-based systems. To fulfill those contracts, they developed their emerging identities as aerospace contractors to promote their projects and recruit additional talent. The Lincoln Laboratory worked closely with IBM to develop the Apollo computer systems, a story well told in David Mindell's *Digital Apollo: Human and Machine in Spaceflight*. Burroughs delivered its second ever B 5000 computer to NASA in 1963—the Burroughs advertisement in *fig.*

*Two very similar-but-different recruitment advertisements. Each has been drawn to attract the interest of "computer people" and makes a graphic case for ground-based support systems operated by skilled engineers.*

Fig. 8.7 (right):
MARQUARDT, WHICH HAD RECENTLY PURCHASED AN IBM MACHINE. ARTIST UNKNOWN BUT LIKELY TO HAVE BEEN KEN SMITH, WHO CREATED NUMEROUS OTHER PEN-AND-INK WORKS FOR THIS COMPANY (*PROC IEEE*, 1963).

Fig. 8.8 (opposite):
RED GATES FOR THE MARTIN COMPANY. THE PROVENANCE OF THE SPACECRAFT DEPICTED IS ANYONE'S GUESS—MOST LIKELY GATES'S IMAGINATION (*SCIENTIFIC AMERICAN*, 1961).

# Marquardt needs electronic engineers

**Involvement: Space Tug** . . . another advanced program at Martin demanding superior talent to explore environmental adjustments for men in space. To individuals seeking the freedom to inquire into such programs, Martin of Denver offers exciting rewards. Write direct to N. M. Pagan, Director of Technical & Scientific Staffing, The Martin Company, (Dept. A-1), P. O. Box 179, Denver 1, Colorado

## Large Scale Computers Speed Engineering and Astro-Navigational Data Processing For Coming Interplanetary Travel!

*7.2* expresses the company's plans in that direction—a computer that NASA then used to coordinate launches of Atlas rockets, including Project Gemini, throughout the mid-1960s.

**PROGRAMMATIC OBJECTIVES OF SPACE ELECTRONICS** NASA and its contractors demanded the rapid development of many electronic systems. In its special issue on space electronics in April 1960 (pictured in the introduction, within *fig. i.12*), the *Proceedings of the IRE* identified additional categories within space applications, in addition to telemetry, communications, and navigations: instrumentation, electronic propulsion in space, and tools for the study of the space environment and its effect on equipment and living things, including people.[15] Implicit in this catalog of objectives is a requirement that all of the systems be developed to function in the demanding environment of space, and adhere to the constraints on size, weight, and cost inherent in launched craft. For industries new to miniaturization and high-altitude environments, these considerations demanded considerable investment in research and development.

**CULTURAL OBJECTIVES OF SPACE ELECTRONICS** The times also demanded that technology define itself in an increasingly anthropocentric environment. This was a significant shift from the wartime and postwar motifs that had animated the emergence of electronics. From Herbert Bayer's "Earth bulb" promoting the FM radio vacuum tube in 1942 (*fig. 1.1*) to Jacqueline Casey's eight-column collage promoting radar astronomy in 1963 (*fig. 7.15*), commercial artists had developed a cultural frame around electronic technologies that positioned them at the center of their own narrative. The artists whose works appear in this book built a corpus of graphic art spanning two decades that mobilized the significance of new technologies, while also eschewing "heroes" in any kind of human sense. New technologies were introduced and contextualized as features, even agents of transformation for human society, nodes in a new network that dramatically expanded the human sensorium. They were the "brain, nerves, and senses of flight vehicles."[16] Where they were pulled by fantasy toward human form, that form was robotic—a play at technology becoming human itself.

This process of contextualization unfolded along two main narrative trajectories: the transit through realms of signification from stylized but otherwise direct representation to abstract, connotative representation, following the course that dominated fine art during the same years, coupled with a geographic storyline that carried the atom from Rutherford's planetary model to the solar system itself.

**ANTHROPOCENTRISM** When NASA introduced the Mercury Seven astronauts to the public in the spring of 1959, it was an abrupt development in the cultural history of technological emergence. Suddenly real people, exactly seven of them, were standing as public emblems of the technological future. Positioning real people at the tip

America's FIRST man into space will rely on a Honeywell designed and developed Attitude Stabilization and Control System for controlling his space capsule. This system automatically damps out initial launch rates, orients and maintains the capsule in proper orbital plane, and provides for the correct descent trajectory and re-entry angle. This device is just one of the many contributions being made by Honeywell scientists and engineers to our nation's space programs.

**Honeywell**

Ⓗ *Military Products Group*

*Fig. 8.10:*
... AND SMALL COMPUTERS: HONEYWELL PROMOTES ITS CONTRIBUTIONS TO ONBOARD COMPUTING FOR PROJECT MERCURY (*MISSILES AND ROCKETS*, 1961).

**flight control for all environments**

of a nosecone, physically at the furthest reach of new technology, complicated the more subtle narratives that had dominated the weaving together of the atom, the planet, and the tube. Astronauts were not allowed to appear in person in commercial advertising, but they did not need to: from the moment of their introduction, the prospect of "man in space" became a heroic motif that permeated much of American culture, from popular entertainment to straight news reporting and everything in between. In industrial advertising, the prospect of human spaceflight inspired legions of stylized feet and hands that sought to reposition the human *step, reach,* and *touch* into space as new heroic gestures[17]—functions that instrumentation alone could not perform.

Within the scientific community the prospect of human spaceflight faced profound challenges of legitimacy,[18] but such objections—that human spaceflight was an unscientific waste of resources—took place away from public view. The figure of the heroic astronaut skewed, even distorted, the public understanding of science and technology. A heroic narrative that privileged direct human engagement over the extended sensory realm made possible by technology invited a pushback from the people making that technology. This was the broader cultural context within which electronics firms continued to advertise their wares. The result is a kind of negotiated truce with the incursion of human beings into the space environment: when promoting robotic advances, technology appropriates, even spoofs, the heroic role, while at other times it strives to depict its own essential "supporting" role in human spaceflight programs.

**SPACE TELECOMMUNI-CATIONS**

The most "natural" extension into space for electronics firms was the continuation of their original purposes: furthering the human senses of sight and sound, and the process of human communication. AT&T's Bell Laboratories had been working toward telecommunications satellites since the mid-1950s, when its lab director, the theorist extraordinaire John R. Pierce, wrote a proposal around the idea.[20] The Laboratories' central purpose, articulated as far back as 1909, was, after all, to transmit a telephone signal over the furthest possible distance within the U.S., coast to coast (*fig. i.3*). Satellite telecommunication was a logical extension of this mission. By 1960 the lab was collaborating with Remington Rand–UNIVAC on NASA's Project Echo, the first telecommunications satellite. *Echo I* and *Echo II* were both passive satellites, basically just big Mylar balloons that transmitted signals

Apollo crews
fly make-believe
by hybrid computer

*Fig. 8.12:*
ASTRODATA. THE COMPANY CONTRACTED WITH NASA TO CREATE CONTROL SYSTEMS FOR THE LAUNCH SUPPORT DIVISION OF PROJECT APOLLO. A DISTINCTLY UNHEROIC ASTRONAUT SPINS IN THE BALANCE BETWEEN A CIRCUIT BOARD AND THE LUNAR SURFACE (*ELECTRONICS*, 1965).

through reflection. Bell Laboratories built the navigation system that calculated *Echo*'s launch-based flight trajectory, while Remington Rand–UNIVAC built the ground-based computer that ran the system.[21]

The laboratory continued to work toward a more ambitious project, the first active telecommunications satellite, *Telstar*. *Telstar* launched in July 1962, and was the first true international telecommunications satellite as we know them today, its transmission capabilities shared with the United Kingdom and France. *Telstar* stood alone as a technological contribution to human society, and as such was commemorated worldwide in song and other popular media. It enjoyed a cultural identity quite distinct from the otherwise bright glow surrounding human spaceflight programs. It was heavily promoted in industrial advertising by the Bell Laboratories subcontractor Radiation, Inc. in a tongue-in-cheek heroic mode (*fig. 8.13*). Note that the artwork in *fig. 8.13* does not assume any familiarity with satellite telecommunications: it spells out the new concept in easily decodable, albeit surrealist, terms.

**AT RADIATION, CHALLENGE IS OPPORTUNITY**

*Example: Bell System's TELSTAR*

Fig. 8.13:
PAUL CALLE FOR RADIATION,
INC: HEROIC *TELSTAR*
(*SCIENTIFIC AMERICAN*, 1962).

**PULSE CODE MODULATION**

Bell Laboratories' pulse code modulation (PCM) technique for long-distance voice communication transmission was well suited for Earth-to-space telemetry systems. Telemetry is the long-distance transmission of coded information from one computer to another; in the case of Earth-to-space communications, telemetry systems enabled ground-based computers to communicate with their corresponding units on board the spacecraft. The collaboration between Bell Laboratories and Radiation, Inc., followed from Claude Shannon's formulation of PCM. Its practical application was a system for the *Telstar* satellite that weighed "only" eight pounds and could carry 112 information channels (*fig. 8.14*). The laboratory cited the Telstar project as an experiment that would make use of a wide assortment of other "ready" electronics components that the lab had on hand: transistors, diodes, antennas, traveling-wave tubes to enhance radio signals, and FM receivers.[22]

The PCM telemetry systems that Radiation Inc. and Bell Labs developed for *Echo* and *Telstar* proved themselves. On the basis of their success, Radiation was awarded the contract to build the telemetry system for the Apollo lunar modules. *Fig. 8.15* is a third recruitment advertisement for Radiation with artwork by Paul Calle. Calle was a noted illustrator of children's books and art instruction books, whose connection to space technology would develop far beyond the series for Radiation, Inc. shown here. In 1963 Calle became one of the first artists invited by NASA to document and interpret the work of astronauts, part of what would become the NASA Artist-in-Residence program. He served a residency at NASA's Kennedy Spaceflight Center, spending time with astronauts and documenting their preparedness training, and would go on to create the U.S. Post Office commemorative stamp issued in honor of the Apollo 11 moon landing of 1969. His artwork for Radiation in *figs. 8-13*, *8-14*, and *8-15*, however, shows a distinct bias toward the role played by electronic engineering relative to that of the astronaut.

**AT RADIATION, IDEAS BECOME REALITY**
*Example: Telstar's 8 lb/112 channel PCM telemetry system*

Fig. 8.14:
PAUL CALLE (PRESUMED) FOR RADIATION, INC. THE HEROIC HAND OF THE ENGINEER . . . AND *TELSTAR* (*SCIENTIFIC AMERICAN*, 1962).

**CRTS AND CIRCUIT BOARDS (RCA)**

The Cold War espionage market offered excellent opportunities for RCA, whose tube division grew from being the new "eye" of television to becoming the supreme "eye in the sky" of early espionage satellites.[23] RCA in the early 1950s began to develop cameras that would be able to monitor Earth from space. This research was prompted by the arrival in the U.S. of German intercontinental ballistic missile engineers, whose skills in

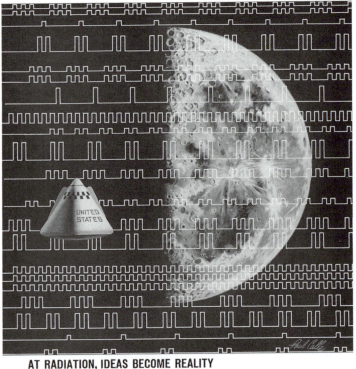

**AT RADIATION, IDEAS BECOME REALITY**

*Example: APOLLO's PCM Telemetry Systems*

rocket and missile development suddenly made the idea of a future satellite age more realistic.[24] The same technology served civil science: RCA's TIROS satellite started out as a spy satellite but was later transferred from ARPA to NASA and served a longer tenure for that agency as a weather satellite. RCA also made computers, having been part of the same process of computer industry assembly engaged in by Burroughs and IBM. The company's model 110 computer was purchased by NASA in 1961 and linked to the system that IBM had built to coordinate operations and rehearse flights by the Saturn rocket.[25] The custom communications circuits that RCA built for this project were doubtless the inspiration for the artwork in *fig. 8.16.*

**DIGITAL NASA** One irony behind the electronics industry's tremendous push toward self-redefinition is that NASA was nevertheless dissatisfied with its progress toward space between 1957 and 1963. Indeed, it wasn't good enough: the loss of all of the Ranger spacecraft to computer and navigational failures, and the losses of many early launches of Atlas rockets, were among the scenarios that contributed to a conviction within NASA that industry was not up to the job. In 1963 NASA called upon Congress to fund its own electronics research center, noting that "available electronics equipment does not meet the needs of many current and future space programs. Most equipment available today is an outgrowth of commercial or military technology which was predicated on different applications in a much less severe environment than space. . . . It is well known that the lack of reliability of electronics equipment has constituted a major factor in launch delays, in-flight failures, and mission terminations."[26] Congress ultimately funded NASA's Electronic Research Center, which was built, at NASA's request, in Cambridge, Massachusetts, near MIT, Harvard, and the Lincoln Laboratory.[27] However, government support for the center waned, and it was closed

ANOTHER ASPECT OF THE MARQUARDT MISSION

# ELECTRIC PROPULSION

*Fig. 8.17:*
KEN SMITH FOR MARQUARDT
(*AVIATION WEEK*, 1961).

in early 1970 after fewer than six years of operation.[28] The ball bounced back to industry to carry space electronics forward.

**ELECTRIC
PROPULSION**
The application of electronics to space exploration that had the longest germination was electric propulsion, based on the energy generated when electrical currents are trapped, compressed, and brought into contact with one another. The several possible electrostatic "electric" motors (electronic devices that are known today as ion drives or ion thrusters) that could power spacecraft were initially developed in the 1950s, at Republic Aviation.[29] The first one finally flew, in 1998, in NASA's Dawn spacecraft.[30] The delay can be attributed to two causes: one, electric propulsion has no dual purpose, no particular military application that would have gained funding for its research during

the Cold War (electrostatic propulsion is not powerful enough to combat G forces, making its central application orbital and interplanetary impulse power), and second, the power supply systems for the engines were unreliable and required considerable investment before they could be proven. This meant that during its development phase electric propulsion had to answer to cultural parameters, even though its pure science application was self-evident. Once the technology was finally proven, those boundaries fell away and the technology has been flying ever since.

The Marquardt advertisement in *fig. 8.17*, supposedly recruiting engineers to work on an electric propulsion project, expresses another odd angle to space-age industrial advertising: companies often placed advertisements, and commissioned artwork to be made, with regard to projects and technologies tangential to their central work, typically with an aim of cultivating their image as vanguard operators. Marquardt (1944-73) was a jet engine company that originally specialized in air-breathing jets, or ramjets, and also in army missiles. In the late 1950s the company branched out into space work and missile-control computer systems (the Marqatron, see *figs. 2.10* and *7.8*). In the late 1950s Marquardt gained a contract from NASA to build the body of the NERVA rocket (Nuclear Engine for Rocket Vehicle Application) for Project Rover.

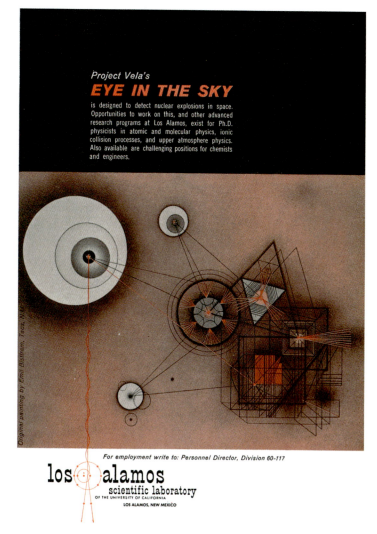

*Fig. 8.18:*

ARTWORK BY EMIL BISTTRAM AS USED BY LOS ALAMOS SCIENTIFIC LABORATORY (*NUCLEONICS*, 1960).

The assignment would doom the company when Rover was canceled in 1972. Their recruitment ad featuring promised opportunities in electric propulsion reflected laboratory research which the company was conducting into hybrid nuclear-ion propulsion systems—systems that have remained in the firmament of science fiction.

**PROJECT VELA**   In the aftermath of the atomic detonations of World War II, there was widespread motivation within both science and government to convert nuclear knowledge to civil purposes. Project Rover's NERVA rocket was one of the U.S.'s three civil nuclear rocket programs of the late 1950s and 1960s. In the postwar years, the Los Alamos Scientific Laboratory (now National Laboratory) devoted one-third of its workflow to the civil uses of atomic science, largely the civil exploration of space. The laboratory researched and developed the nuclear propulsion system for NERVA, which was one of its largest projects of the 1960s. The nuclear weapons research, development, and maintenance for which the laboratory is best known today became its essentially full-time duty only after the 1972 cancelation of NERVA. In addition to the nuclear propulsion system, the lab's scope of involvement with civil space sciences in the 1960s was considerable, and included a wide range of atomic science-based investigations. It created the gamma-ray spectrometer that the ill-fated *Ranger 3* carried past the moon, which returned useful results in spite of the craft missing its target. Another of its pursuits was Project Vela, designed to verify compliance with the Limited Nuclear Test Ban Treaty of 1963.[31]

Project Vela comprised both ground-based and satellite-based sensing systems to detect nuclear detonations in the upper atmosphere.[32] The advertising and recruitment campaign conducted by the laboratory to bring qualified scientists and electronic engineers to work on its projects was an astonishing combination of art and science.[33] During the interwar years the New Mexico desert was a destination for European émigré painters of the Expressionist school as well as painters from throughout the U.S. and elsewhere. Many of these artists were deeply inspired by the landscape and developed a style that was specific to the region's stark collision of desert and sky, and which also drew on Native American textile design motifs for inspiration (*fig. 8.18*). (The Shiprock, New Mexico, site where Fairchild would build its semiconductor plant is within one hundred miles of Santa Fe, the hub of the midcentury New Mexico desert fine arts movement.)

The painter Emil Bisttram began a series of paintings called *Space Images* as early as 1954.[34] In an arrangement about which more remains to be learned, the personnel director of the Los Alamos laboratory discovered these paintings and developed them (or appropriated them?) into the basis of a multiyear recruitment campaign.[35] The resulting series of magazine advertisements was published between 1959 and the mid-1960s, and was unlike any other ad campaign; uniquely, it featured artwork that had originated as fine art and that, presumably (being shown in galleries, collected by collectors), maintained its identity as fine art even after being used commercially.

**THE NEW-ELECTRONIC BODY**   Space electronics defied the motif of the heroic astronaut in many ways, not least by becoming the nerve system of a new "body." When we launched satellites into orbit, our tools formed a new ring around Earth. Our space-based installations are governed by the same three physical laws that hold atoms, planets, and solar systems together, hence our off-Earth objects are bound together in durable rings. They form a system of things, both networked and free-floating (interplanetary spacecraft). Electronic telecommunications networks are the nerve system of what has become a space-based corpus—or body—of technological objects. Satellites have allowed us to extend our human sensorium to a degree unimaginable without electronics: satellite-based deep space telescopes look outward, while telecommunications technologies facilitate global telepresence. Within that "body," the accumulation of space junk since the dawn of the satellite age is a virus that keeps spreading. Our next "frontier" may be to reconcile this new "body" with the expectations of care and longevity that we lavish on our own health, and to hold it accountable for the new realms of ethical and social conflict that it has enabled. The UN's Outer Space Treaty of 1967, intended to preserve the use of space for peaceful purposes, was developed in response to Cold War saber-rattling, as the USSR and the U.S. were both conducting nuclear detonations in the atmosphere in the early 1960s. Our contemporary use of satellites for everything from espionage to drone war is an end-run around that restriction. In the meantime, passive neglect of retired satellites has led to severe pollution of the orbital environment, a situation that is coming to threaten all uses of space.

In 1976, Carl Sagan, Ann Druyan, and their collaborators created the *Voyager* disk, an artist- and scientist-made work that dared to depict planet Earth to the cosmos in general terms. Included in the cultural artifacts encoded on the disk, among recordings of music and terrestrial soundscapes, was a graphic image depicting a distinctly European pair of human beings in sketch outline form. The disk was the earliest artistic intervention that reached directly into the newly accessible landscape of the solar system, an artwork among the "body" of human-made objects in space. Launched on the *Voyager* space probe on a trajectory that carried it outward, away from the sun, it entered the outer limits of the solar system in 2013 and is currently on its way into galactic space.

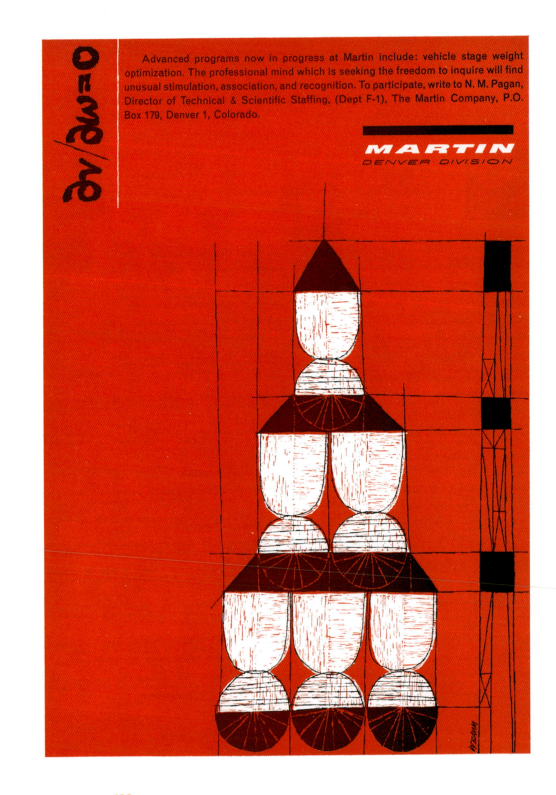

In 2012 the artist Trevor Paglen placed a curated selection of one hundred archival photographs into a satellite and arranged to have it launched into a stationary orbit around Earth. Each of the photographs offers a perspective on the impact that human beings have had on the planet, showing shaping forces and natural and cultural phenomena of all kinds. The project, titled *The Last Pictures*, is a slice of life from the Anthropocene. It was developed for launch in collaboration with materials scientists at MIT (the photographs were microscopically etched into a disk and then gold-plated). On board the telecommunications satellite *EchoStar XVI*, the photographs will orbit Earth for millions of years, long after *EchoStar* has outlived its usefulness and migrated to a graveyard orbit (above and outside of active orbital devices). Its longevity also guarantees that it will orbit Earth long after the tenure of human life on the planet will likely have expired. Paglen's essentially permanent project is akin to putting a tattoo on a distant part of this new electronic "body." It is an early artist's attempt to weave artistic annotation into the landscape of space hardware. For all that anyone might think about Paglen's project itself, it is perhaps most interesting as a provocation: What else might artists do with this newly made body?

*(opposite) Fig. 8.19:* WILLI BAUM FOR THE MARTIN COMPANY. THIS IS ARTWORK FOR ENGINEERS: MATHEMATICAL PROPORTIONS ANCHOR A STYLIZED ROCKET, INTRODUCING AN ARCHITECTURAL SENSIBILITY THAT INVOKES THE DESIGN OF A JAPANESE PAGODA [*MISSILES AND ROCKETS*, 1961].

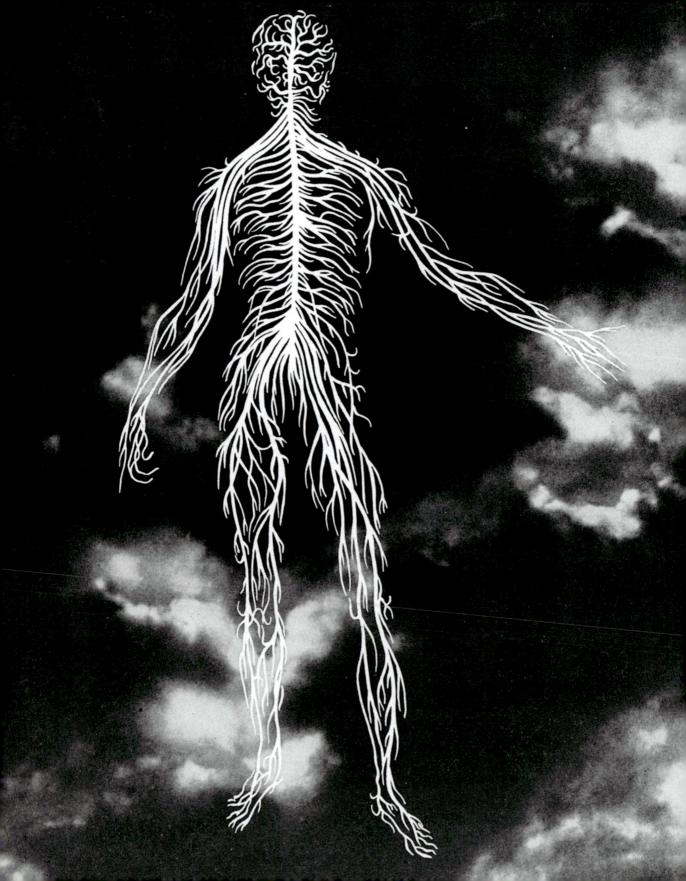

## CHAPTER NINE

# BIONICS, A PROLOGUE TO TRANSHUMANISM

An astronaut reenters Earth's atmosphere during the test flight of an experimental space plane, Northrop Aircraft's M2-F2. NASA footage of the spacecraft's spectacular crash landing at Dryden Spaceflight Center anchors the opening scene, lending an unusual realism to this science fiction drama. The former moonwalker survives the crash to embark on the next frontier to follow his lunar explorations: the integration of the remnants of his body with electronic components. The components make him "better, stronger, faster" than he was before. He is now a cyborg.

> —summary of the opening sequence of *The Six Million Dollar Man* (series pilot), a television show based on the 1972 novel *Cyborg* by Martin Caidin[1]

The novel *Cyborg* by the space technology writer Martin Caidin is both good science fiction and a critical metanarrative that maps the cultural turn toward the body and away from outer space. Caidin's narrative proposes that the astronaut's next frontier, following the conclusion of the Apollo program, was a new life as a cyborg. Published the same month as the last moon landing (December 1972), Caidin's novel was an intervention into the publicly embraced motif of the heroic astronaut. With the mass popularity of the TV adaptation, *The Six Million Dollar Man*, the novel's metanarrative entered mainstream American culture, helping to redirect the public's attention away from space and toward "bionics" as a harbinger of the technological future.

The human spaceflight program did not end in 1972, but it did pause

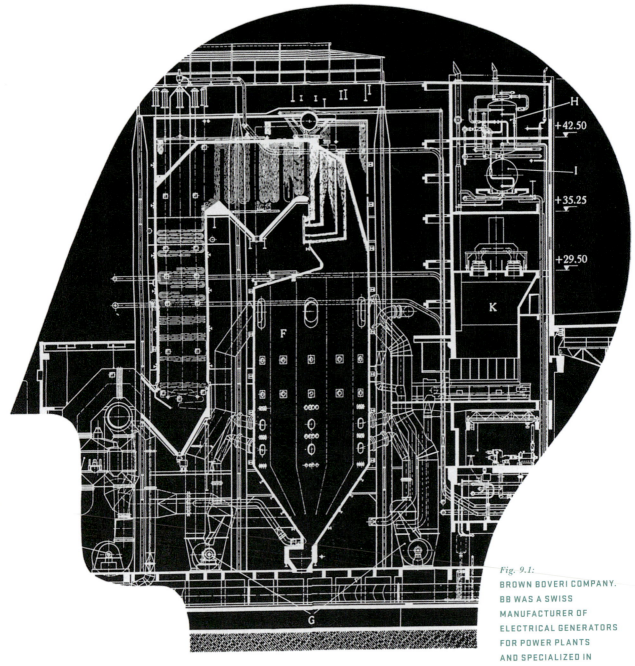

Fig. 9.1:
BROWN BOVERI COMPANY.
BB WAS A SWISS
MANUFACTURER OF
ELECTRICAL GENERATORS
FOR POWER PLANTS
AND SPECIALIZED IN
CONTRACTS FOR THE U.S.
CIVIL NUCLEAR INDUSTRY
DURING THE COLD WAR
(*ELECTRICAL WORLD*, 1962).

+42.50

+35.25

+29.50

and reorganize. While it was doing so, robotic spaceflight ramped up. In 1976 not only was the *Voyager* probe launched toward the edge of the solar system but the *Viking* landers commenced their Mars missions, sending the first photographs back to Earth showing what the surface of the red planet "really" looks like. Human spaceflight was meanwhile reoriented in 1980 to a research-based presence in low Earth orbit. In the years since, robotic exploration of outer space remains among the most exciting and the most scientifically and humanly meaningful applications of space technology.

But for advanced electronics, the human body has emitted a denser gravitational field. The dual spatial tension between the atom and the planet that opened this book reconstitutes as a triad, joined by the far more personal sphere of the human body. Today, for instance, a leading edge of circuit design is focused on circuits that can be integrated into living tissue for medical purposes, dissolving inside the body after use.[2]

For as long as electricity has been cast as a doppelganger for life itself, electric phenomena and electronic devices have been understood as extensions of organic human abilities. In earlier chapters devices such as radio, television, and radar were framed as extensions of the human sensorium, introducing the close relationship between electronics and the human body. The advertising art in this chapter engages the human body as a place where the two domains of human and machine are at work adjusting each other's very definition. This chapter therefore has the longest chronological trajectory in the book, reaching back to the emergence of the robot in the 1920s and reaching ahead to touch today. The preceding chapters have been a series of interwoven chronologies with the close of each chapter bringing us a few years further into the future. The journey here from the vacuum tube era to "bionics" concludes this series, leaving us at the doorway of the world we are living in.

The graphic art in this chapter invokes the changing boundaries of our humanness, art that was created first to introduce the emergence of advanced computing and, later, applied bioelectronic technologies. This art is made within a cultural context as big as that encompassed by the narrative art of science fiction, to which it bears a close relationship. The robot figure, a "character" borrowed from science fiction, is a machine that resembles a human being. Its mirror opposite is a cyborg, a human being (or other living thing) that has had its nature changed through the addition of electronic parts.

These two poles of convergence between human and machine are reciprocal and dialectical in nature. It is within this dialectic that many

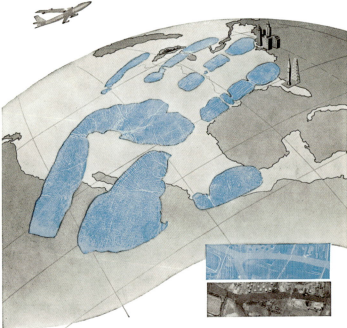

**INFRARED FINGERPRINTS**

for reconnaissance, interpretation and prediction

changes took place within the field of electronics in the mid-twentieth century. New developments within the field, both technological and cultural, were named and interpreted based on their perceived, or assigned, position between the poles of biological and mechanical. The language of electronics evolved within this dialogic loop, as did the identity of the field itself. Accompanying these processes was art made for industry, both graphic art and art in other media. A majority of the world's science fiction literature has been incubated within this dialogic space as well, and to a greater extent than any other chapter it forms a backdrop for the artwork explored here. At the societal level, these processes have inspired large subdisciplines of interpretive works. This particular chapter acts as a short prologue to the ongoing project of mapping the permeable boundary between human and machine.

**ROBOTS APPEAR IN INDUSTRIAL ADVERTISING**

As in space sciences, the narrative arts anticipated and inspired the work of scientists and engineers for decades leading up to the development of new technologies. Shelley's *Frankenstein* wrapped a laboratory science environment around an otherwise mythic breach between death and life. A short century later, one of the first mechanical characters in popular literature was L. Frank Baum's Tik-tok, the mechanical man, introduced in the novel *Ozma of Oz* in 1907. When the word "robot" entered global languages in 1921 through the work of the Czech playwright Karel Čapek, the idea of a wholly artificial, mechanical or electromechanical "person" became more fully formed. Crafters of science fiction developed this idea in the decade that followed, building one scenario after the next to explore the possibilities opened up by the emergence of robots into human society.

The appearance of robots in feature films, such as Fritz Lang's 1927

*Metropolis,* established the chiseled (machine-made) facial appearance of the prototypical robot and its characteristic passive facial expression. Robots appeared as spokespeople for industry as early as 1928. To celebrate its development in 1924 of Televox, that switching system that was really a rudimentary form of artificial intelligence, Westinghouse built its first promotional robot, Herbert Televox, which debuted at fairs and other events in 1928.[3] Because its vacuum tubes were so hot, this early robot featured an open chest cavity with its electronics exposed to cooling air.

Two years later another robot figure appeared in a long-running printed ad campaign on the pages of the *Literary Digest* (*fig. 9.3*). Sponsored by the American Federation of Musicians (AFM), the ad series critiqued automation, expressing the union's strident opposition to "canned music"—player pianos and phonographs—in theaters. This was an early and unusual negative use of a robot character, in this case in a campaign by labor to draw attention to a threat to working artists. It is also unusual in its origins. It was drawn by Jesús Helguera, the Mexican-Spanish painter best known for painting heroic mythic and historical figures from Mexican history. As a young art student in Spain between the wars, Helguera created commercial art for magazines to help fund his education and is responsible for the AFM series.

*Fig. 9.3:*
JESÚS HELGUERA FOR AMERICAN FEDERATION OF MUSICIANS (*LITERARY DIGEST,* 1930).

# A PICTURE NO ROBOT CAN PAINT!

*Fig. 9.4:*
RCA'S "MAGIC BRAIN" (1934).
FROM THE SLEEVE OF A 78RPM
RECORD.

Science fiction reached its first phase of maturity in the 1930s, and lines between fiction and reality were actively blurred in that decade, most notably by Hugo Gernsback. Both a radio engineer and a science fiction promoter, Gernsback simultaneously published fiction magazines and many "straight" trade periodicals including *Radio-Electronics* (founded in 1929 as *Radio Craft*). He actively discouraged sharp distinctions between the two.[4] Throughout his career, Gernsback promoted a vision of technological development in which science fiction was its equal and necessary partner, a dream state without which the "waking" state of technology could not move forward. The literature of the 1930s that Gernsback promoted was populated with robots, and the concomitant anthropomorphization of new technology increased in visibility around the same time.

In 1934 RCA introduced the "Magic Brain" (*fig. 9.4*), an electronic signal processor that was integrated into both radio receivers and phonograph arms to enhance the sound quality of the audio signal.[5] It was widely promoted in 1935 with a graphic motif that appeared everywhere from magazine spreads to the dust sleeves of RCA long-playing records. In this motif a human head is cut away to reveal an electronic "brain." The figure's quasi-human head evokes an early logo of the Bauhaus school, a similarly linear facial silhouette. By the 1930s the figure's closed, passive, mechanically sculpted face expressed not just modern design, but roboticism.

1935 was an eventful period of convergence between laboratory electronics and the human body: *Electronics* magazine reported in that year that *thinking* had been discovered to be an electrical process. This news followed centuries of inquiry into the effect of electricity on the body, including the nineteenth-century experiments with x-rays that facilitated discovery of cathode rays. *Electronics* placed due emphasis on the nature of the relationship between electronics-based investigation and the human body:

> Electronic amplifiers now reveal that the brain is also the seat
> of teeming electrical currents which flow back and forth in
> the mysterious process called thinking. . . . amplifiers, in the

form of the cardiograph, have shown that electrical flashes accompany each movement of any muscle. The heart muscles generate an EMF of a millivolt or so. . . . With the human body itself now recognized as being but a "bundle of electrons," it is not surprising that the electronic amplifier is proving the master tool of this new assault on the mystery of life.[6]

The electronic microphone was also developed that year, replacing the mechanical microphone and Victrola-style horn in the role of capturing sound for recording, bringing the depth and intensity of recorded sound one step further into people's ears.

Three-dimensional, electromechanical robots kept up their work as well. In 1934 the Franklin Institute in Philadelphia (the museum dedicated to the work of Benjamin Franklin) activated a robot greeter named Egbert at the museum's entrance.[7] Egbert was operated by photoelectric switches set up so that opening the door to the museum triggered an electromechanical arm to be raised in greeting, and a phonograph to play a recorded welcome message. One year after Egbert, Westinghouse built the robot Elektro to act as a "spokesperson," communicating concepts of automation to the public. Elektro was much more advanced than his predecessor Herbert Televox. When presented by Westinghouse at the 1939 New York World's Fair, Elektro was a cigarette-smoking, brainpower-bragging aesthetic sculpture, six and a half feet of sleek design.[8] His chiseled appearance promoted Westinghouse's capacities in data processing and electronic and electromechanical switches of all kinds. Westinghouse was also the manufacturer of the first portable electrocardiograph, in 1930, so the robot's appearance reinforced the company's role as a maker of biomedical electronic devices.[9]

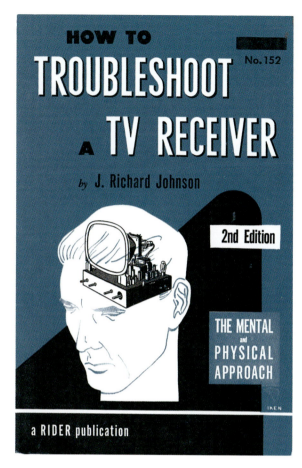

*Fig. 9.5:*
TV ON THE BRAIN: AN INTERESTING GAMBIT TO ENGAGE A TV REPAIRMAN, INVITING [HIM] TO IDENTIFY WITH A ROBOT, OR THE TV ITSELF (TV REPAIR HANDBOOK, 1958).

**SITUATING ROBOTS IN GRAPHIC ART** Technology's influence on the visual arts of the twentieth century found expression across the whole scope of modernist possibilities. Modern art can even

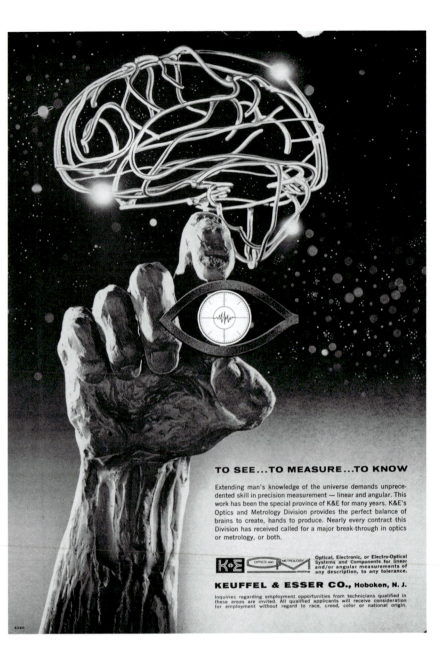

*Fig. 9.6:*
KEUFFEL AND ESSER
COMPANY. BUT IS IT ART? THIS
SURREALIST ADVERTISEMENT
FROM A MAKER OF SLIDE
RULES SUGGESTS SOMETHING
ABOUT THE TENSION BETWEEN
ANALOG AND DIGITAL MODES
OF COMPUTATION, BUT
IT'S HARD TO KNOW WHAT,
EXACTLY (*MISSILES AND
ROCKETS*, 1961).

be defined in part by its engagement with technological (geometric) forms,
and its simultaneous questioning of the centrality of the human figure as a
reference point. Herbert Bayer's work that moved Earth, the tube, and the
sky into the heart of the figure/ground schema can be seen, in this context,
as a nodding wink to these tenets of modernism. Tension between organic

and geometric forms in graphic modernism, as seen in the depiction of crystals, reflected a similar tension in the shapes created by abstract artists. The trajectory of electronic devices from discrete objects toward diffused, networked systems was ultimately mirrored by movements in visual art, primarily in the sphere of fine art, that led away from representation of any kind.

Graphic artists making work to interpret robotics in the context of industry therefore worked within a sphere of representation bounded by four distinct realms. There was occasional influence from modernist fine art traditions (surrealism, *fig. 9.6*), but the very different influence of graphic motifs from science fiction (also *fig. 9.6*, definitely *figs. 9.3*, *9.4*, and *9.5*) was often more relevant to the service of science-based industries. Third was the presence at the intersection of art and industry of nongraphic figures such as kinetic sculpture (three-dimensional robots). Last is the influence of scientific research itself. No less than astronomy or crystals science, the combination of electronics and biology relies on rigorous, long-term laboratory research.

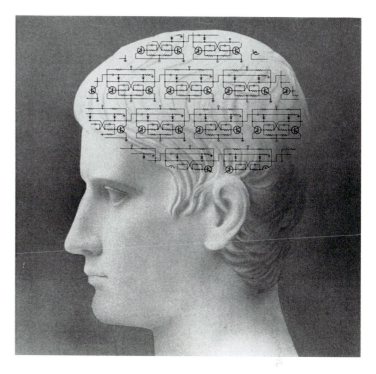

Fig. 9.7:
GILFILLAN (*AVIATION WEEK*, 1959).

Fig. 9.8:
"SON OF MAGIC BRAIN": DETAIL OF THE JACKET ART FOR EDMUND C. BERKELEY'S *GIANT BRAINS, OR, MACHINES THAT THINK* (1949).

## CYBERNETICS, AND GREAT BRAINS

During the great laboratory decade of the 1940s, the work of biological researchers combined with the work of theoretical mathematicians. The English mathematician Alan Turing developed a theory of artificial intelligence that combined psychology with computational mathematics to form a new way of looking at cognition. At the same time, Claude Shannon's work on

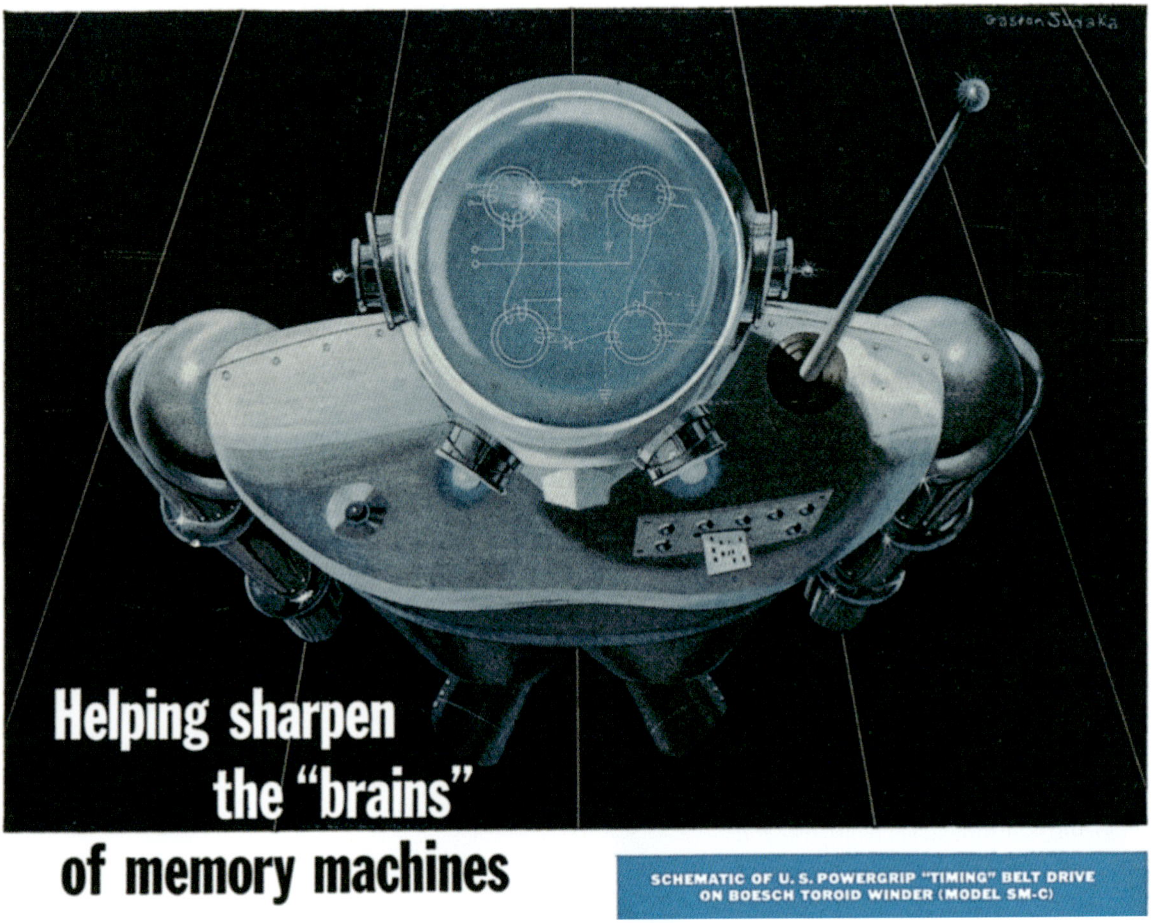

Helping sharpen
the "brains"
of memory machines

SCHEMATIC OF U. S. POWERGRIP "TIMING" BELT DRIVE
ON BOESCH TOROID WINDER (MODEL SM-C)

*Fig. 9.9:*
GASTON SUDAKA FOR
UNITED STATES RUBBER. AD
PROMOTING THE COMPANY'S
WORK IN SUPPLYING THE
COMPUTER INDUSTRY WITH
THE TIMING BELTS NEEDED
BY MAINFRAME COMPUTERS
(*MISSILES AND ROCKETS*,
1957).

communication theory at Bell Laboratories was inspiring disciplinary developments in information theory.

Following several years of transatlantic research and transdisciplinary
collaborative investigations, the newly named field of cybernetics coalesced
in 1948. It was fully laid out in a book of that title by the logician, philosopher,
and mathematician Norbert Wiener.[10] *Cybernetics* was a landmark attempt
to define the emerging domain of unified inquiry that joined biological communications research to logical-mathematical theories of communication,
taking both information theory and artificial intelligence into account. Wiener was one of a large international group of mathematicians and computer
scientists—including Shannon and John von Neumann—whose collaborative
work and conversations in the 1940s and '50s yielded the ur-narratives that

permanently shaped the aspirations of computing. Wiener's work on cybernetics, however, expanded beyond its influences to describe a system of networked reciprocity between technology and society. In Wiener's view, human behavior and computer system behavior were reconcilable and complementary, and when combined would generate a new society that would be specifically "homeostatic"—democratic and self-regulating.[11] He was writing during the aftermath of World War II, and his philosophy sought to draw a new horizon for humanity partly in response to that crisis.[12]

Cybernetics is a philosophy, not a technology, but information theory and artificial intelligence theory are its intellectual progenitors. Those closely linked disciplines were not immediately applicable to industry and commerce in the way other 1940s' inventions like the transistor were. However, they had a dense and fruitful second laboratory decade during the 1950s. Together, cybernetics, information theory, and AI theory drove a tidal wave of pure research during the early Cold War. The burgeoning surveillance industry, which was a focal point of war production during that time, was a "natural" domain for experimentation in these fields: the prospect of harnessing animal sense perception to the development of advanced, experimental circuits and sensor devices was highly attractive to well-funded military interests. These disciplines already shared common origins within World War II-era military contract work, and the laboratory research in these areas remained funded in large part by the army and navy throughout the Cold War. Yet at the same time, the 1950s was a time of flourishing internationalism within the field. Soviet computer scientists visited the U.S., and vice versa, in the late 1950s, and the Russian-language journal *Problems of Cybernetics*, inaugurated in 1959, was reprinted in English and distributed in the West.

Relative to commercial electronics, cybernetics worked like science fiction, as a "dream state" that drove a vision of what the future could be. Information theory and AI theory worked more like the solar system toward which the space programs aimed. Artists began to animate the ideas behind them as graphic motifs in the middle of the 1950s, even though the theories took much longer to filter into tools that were used by ordinary people. While they led advances in computer science, their impact on the kind of technologies that were depicted in graphic print advertising was meta-contextual. There were not actually cybernetic robots being made and marketed in the mid-1950s when images of robots begin to appear frequently as graphic elements in advertising. Rather, the idea of a future populated by intelligent machines and life-enhancement tools for human beings defined a technological ideology that informed both the back end (the research and development) and

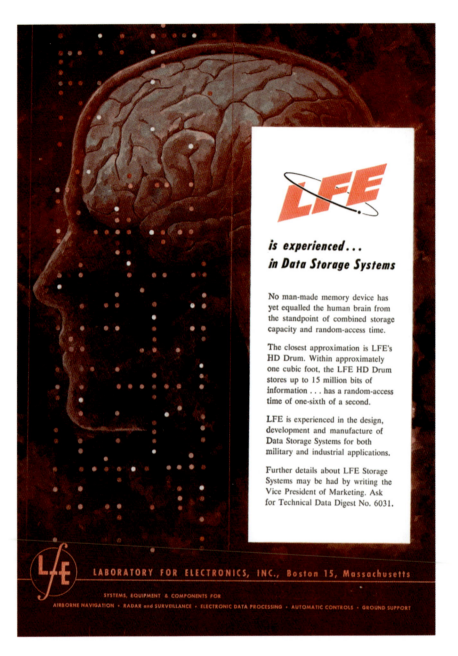

*Fig. 9.10:*
LABORATORY FOR
ELECTRONICS. THE
REMARKABLE PERSISTENCE
OF PAPER TAPE PUNCHED
HOLES (*SCIENTIFIC AMERICAN*,
1960).

the front end (the public-facing representations) of emerging real-world
technologies.

One immediate, short-term effect of the publication of cybernetic theory was the sudden regular conjunction of the terms "brain" and "computer."

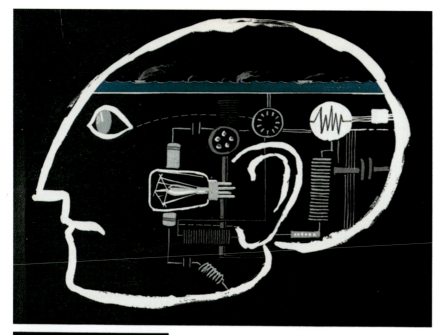

### A sophisticated weapon   The Fairchild Petrel

*Deadly accurate . . . exceptionally reliable*

Bombers armed with the new Fairchild Petrel missile can release the deadly "birds" well outside the protective wall of shipboard AA. Petrel *thinks for itself* . . . seeks out its victim and strikes with uncanny accuracy.

"Petrel performs with a degree of relia-

bility exceptional for such a sophisticated weapon", the Navy says.

Petrel has been fully operational for some time. It is a masterwork of advanced electronics and daring design . . . another major contribution to our national defense . . . by Fairchild.

**FAIRCHILD**

GUIDED MISSILES DIVISION · WYANDANCH, LONG ISLAND, N. Y.

*A Division of Fairchild Engine and Airplane Corporation*

. . . WHERE THE FUTURE IS MEASURED IN LIGHT-YEARS!

*Fig. 9.11:*
FAIRCHILD. AN "ELECTRONIC BRAIN" MOTIF PROMOTES "A SOPHISTICATED WEAPON," THE PETREL MISSILE. THIS PARTICULAR REPRESENTATION UNWITTINGLY ADDS ANOTHER LAYER OF COMPLEXITY UNRELATED TO THE PRESENT SUBJECT: THE NAMING OF THE WEAPON AFTER PETRELS, A FAMILY OF (PEACEFUL) AQUATIC TUBENOSE BIRDS (*AERO-DIGEST*, 1956).

This linguistic phenomenon took place simultaneously in technical and popular literature in the year after the publication of *Cybernetics*, which featured a chapter provocatively titled "Computing Machines and the Nervous System." That year, 1949, the information theorist and AI researcher Warren McCulloch published in *Electronic Engineering* "The Brain as a Computing Machine," a paper exploring the physiology of the brain from an electronic engineering perspective, with implications for computer development. McCulloch was a colleague of Norbert Wiener, and a longtime excavator of the intersection between biological and automatic ways of "thinking." He would go on to influence the development of computer logic systems into the 1960s, and he was a mature member of the community even in 1949. But before Wiener published *Cybernetics*, McCulloch had not yet put his ideas into precisely those terms.

Also in 1949, the computer scientist Edmund C. Berkeley published the mainstream book *Giant Brains, or, Machines that Think*. The book was an effort to drive public understanding of computers (if not actually to *popularize* computers, as none were yet available to the public). Both McCulloch's paper and Berkeley's book cited Wiener's influence, and Berkeley's book promoted *Cybernetics* on the back of the dust jacket. Berkeley's book is actually the first popular explanation of computing, and is also a short history of the new field. Berkeley had founded the Association for Computing Machinery in 1947, and went on to found the journal *The Computing Machinery Field* (1951), soon retitled *Computers and Automation* and later retitled again as *Computers and People*. Berkeley was a computer scientist, not an academic, but his periodicals were not sloppy: for instance, he published Grace Hopper's early essays on her philosophy of developing new compiling routines in 1953.[13]

In 1951 the president of Bell Laboratories, Marvin Kelly, reflected on the invention of the transistor which had taken place three years earlier in the lab. Kelly hypothesized that the transistor's small size and efficiency would "make the telephone system of the future much more like man's brain and nervous system." Kelly's remarks express the rapidly evolving identity of the lab, following the nearly simultaneous, if originally unrelated, developments of the transistor and Claude Shannon's formulation of PCM. World War II was over, and electronics research was recasting its program for decades of Cold Peace, and wars of the future. Given the lab's position at the time as the leading research and development think tank for electronics and affiliated inquiry, Kelly's metaphor had implications beyond the work of his audience members at the Bell System Lecturers' Conference.[14]

The following year the *Proceedings of the IRE* proposed that electronics, as a field, be redefined. Citing the contemporary irrelevancy of the original vacuum tube-based definition ("the science and technology of systems using devices in which electrons flow in a gas"), the editorial, by the engineer and former institute president William L. Everitt, updates the definition to: "Electronics is the science and technology which deals primarily with the supplementing of man's senses and his brain power by devices which collect and process information, transmit it to the point needed, and there either control machines or present the processed information to human beings for their direct use."[15]

Everitt's proposal, following the publication of *Cybernetics* by three years, points out the expanding identity of the electronics field. It opens electronics to be redefined by emerging developments in computing; it opens electronics to allowing particulars of software and hardware to play definitional roles. It anticipates the emergence of networks, in which processed information is "presented . . . to human beings for their direct use." The history of electronics, as it followed from nineteenth-century electrical engineering, had always been centered on tools through which electrons offer a power assist to the human sensorium. The earliest electronic inventions were communications devices as well as extensions of hearing, sight, and sound. Over the twentieth century, the technological developments that facilitated these extensions seemed to point toward a possible future *posthumanism*—a time when the limitations of the human condition would be transcended by technology.

This four-year postwar historical "moment" of 1948-52, within which Wiener's *Cybernetics* was published and Kelly and Everitt made their definitional pronouncements, formed a turning point for the cultural frame of what electronics were and what electronics meant. This window in time corresponds directly—if perhaps coincidentally—to the transistor's four-year cycle of emergence between its invention in 1948 and its implementation in 1952. From that "moment" onward, the theoretical world of artificial intelligence and the fictional proving ground of sci-fi robots were pulled out of their respective cloistered origins to become part of the public context for the development of electronics. This process was not wholly unlike the later shift in the center of gravity of circuit development from radio to computing. The "brain" motif that had started out as a signal processor for audio technology was appropriated from radio and record players to the rather more apt domain of computing. A couple of years later, graphic artists began to utilize "electronic brains" as motifs for conveying all kinds of developments in the field.

**THIS ROBOT WILL NOW REDIRECT YOUR ATTENTION**

The "electronic brain" motif became widespread in the mid-1950s and gained popularity as industry expanded its use of computers. Before long, companies whose primary business did not concern computing were using "electronic brain" images and robot motifs to promote any product that contained even a small computer component—or even those that didn't: mechanical gyroscopes were also promoted as "brains."[16]

In 1958 the Monroe Calculating Machine Company introduced a digital computer named the Monrobot, which quickly became among the dozen most popular midsized computers in the country.[17] Robots, in contrast to successful business computers, were potentially subject to "human" failings. In *fig. 9.12*, a startlingly realistic painting of a very human-looking robot apparently in mid-development appears in a Martin recruitment advertisement. In the original, it is accompanied by an equally startling fictional conversation between a scientist and a potential recruit:

*Recruit:* "You mean that you could actually build a mechanical mind? One that would exhibit emotions——such as love, fear, anger, loyalty?"

*Scientist:* "We're doing something like that now in advanced missile development . . . The 'pilot' that is being developed for the big long-range missile. *He* has a wonderful memory . . . *he* loves perfection, and *he* becomes highly excited when *he* gets off-course . . ." [Emphasis in the original.]

The advertisement dates to 1958, the year that the Martin Company won an army contract to build the Titan, a nuclear warhead–bearing intercontinental ballistic missile. The final, added layer in the jigsaw semiotics of this particular advertisement, however, is the contrast between the planned use of Titan and the actual use of Titan. Given that the U.S. did not engage in the kind of "hot" war that Titan was made for, its hardware was adapted in the 1960s and 1970s to serve the civil space program. Its ultimate value lay in the Gemini missions and numerous robotic missions for which it served as a launch vehicle. Most of Martin's recruitment ads traded on the glamour associated with the civil space program; this one

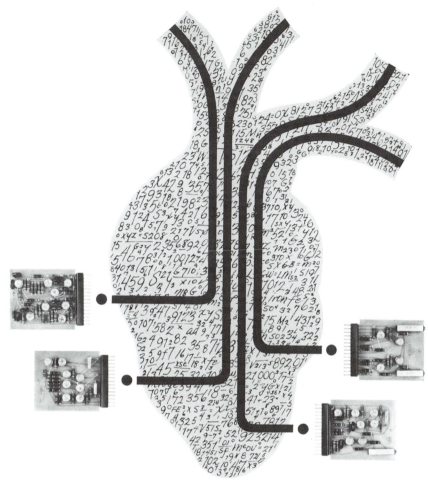

*The pulse of the test range...*

Fig. 9.13:
REDIRECTION AT WORK: IS IT
BIOMEDICAL ELECTRONICS?
IS IT JUST VISIBLE
LANGUAGE? IT'S A SEMIOTIC
JIGSAW PUZZLE IN SERVICE
OF PROMOTING VITRO
LABORATORIES' LINE OF
CIRCUIT BOARDS FOR TEST
RANGE TIMING SYSTEMS
[*ELECTRONICS*, 1962].

stands out. It was obviously designed to attract the kind of mathematical computer scientist who would be caught up in a fictionalized AI scenario turned robot "narrative."

**APPLIED AI: BIONICS** The term "bionics" was coined around 1960 to refer to the recently emerged field of machine processes that mimic biological organisms.[18] Later it came to refer to people who possess machine attributes, a synonym for "cyborg." This usage appeared in the 1970s, as for example with the "bionic man," a nickname for Martin Caidin's science fiction hero. Here I'm restricting its use to its original sense; that is the sense meant by practitioners in the field when they initially

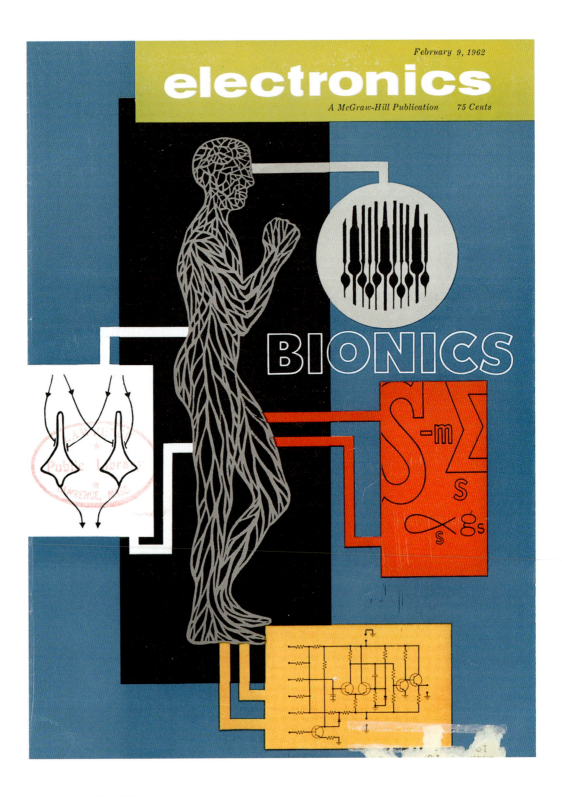

dubbed their discipline,[19] and it's the sense meant by *Electronics* magazine in their four-part 1962 survey "Bionics." In general, the change in signification was a one-way journey: the term first conformed to *Electronics'* 1962 usage and only later was appropriated to refer to reciprocal sciences.

Speech synthesis and computer "reading" were both central to laboratory investigations that took place within the broader rubrics of information theory and artificial intelligence. The Bell Laboratories work on computer speech described in chapter 7 was the most advanced of several comparable investigations, both in England and in the U.S., aimed at producing computers that could mimic human communication behaviors. The other nerve systems of activity were the numerous programs to develop computation systems that could mimic human cognition and memory—literally "neural" computing—and programs to simulate the behavior of every other human sense. These were legion during the Cold War; research institutions with programs of this type include the Rand Corporation, the Carnegie Corporation, Bell Laboratories, the Army Signal Corps, MIT's neurology and biophysics laboratories, and the Cornell Aeronautical Laboratory.

Some 1950s-era research into AI developed independently of the process of digitalization and transistorization that we have watched unfold in the same decade. One project that garnered attention as early as 1958 was Cornell Aeronautical Laboratory's Perceptron, an analog pattern recognition system described as an "artificial nerve network consisting of logically simplified neural elements."[20] The Perceptron was a system of electronically-powered electromechanical relays—vastly more sophisticated than Westinghouse's Televox control system or its Elektro robot, but not fundamentally different from them. Sponsored by both the army and the navy, the Perceptron system was used to model machine intelligence into the 1970s, before its limitations were fully mapped. It's of note here because it's

*(opposite) Fig. 9.14:*
THE COVER OF THE FIRST IN A FOUR-ISSUE *ELECTRONICS* SERIES WITH A SPECIAL FOCUS ON BIONICS (1962).

*Fig. 9.15:*
CORNELL AERONAUTICAL LABORATORY. FROM ROBOT TO . . . UNSETTLINGLY ALERT EYES AND IRREGULAR, "NATURAL-LOOK" FACIAL DETAILS CLUE US THAT THIS IS NO ROBOT; THIS IS ARTIFICIAL INTELLIGENCE. NOTE THE RUTHERFORD-BOHR ATOMIC SYMBOL USED HERE TO CONNOTE INTELLIGENCE (*PROC IRE*, 1958).

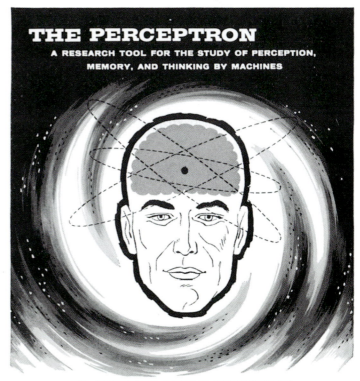

THE PERCEPTRON

A RESEARCH TOOL FOR THE STUDY OF PERCEPTION, MEMORY, AND THINKING BY MACHINES

A NEW FRONTIER
AT CORNELL AERONAUTICAL LABORATORY

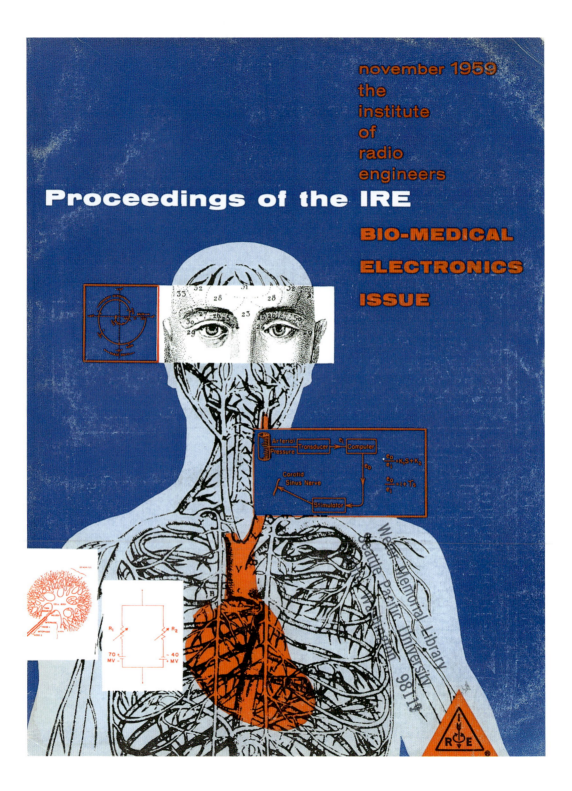

one of the very few projects that published a recruitment advertisement specifying a call for people to work on artificial intelligence. Its significance in the history of technology is centered on the controversy it engendered over the theory and future of AI research between its inventor, Frank Rosenblatt, and his fellow AI futurists Marvin Minsky and Seymour Papert, who sought to disprove the machine's usefulness.[21] The artwork in *fig. 9.15*, created to promote it, interrupts the tradition of blank-faced robots with a creepy, almost-human look that falls into the "uncanny valley" of near-human images that our brains are unable to easily process.

By 1960 AI research had gained stature within both hard science and computing. In the early 1960s it was subject to a new phase of interpretation and integration into both philosophical and applied levels of computer theory. At the philosophical level, a new essay that year by the cyberneticist J. C. R. Licklider, "Man-Computer Symbiosis," proposed that humans and machines would soon converge to form a symbiotic system, one among a range of possible close relationships between people and computers.[22] Licklider noticed that human thought was messy but multivalent, whereas computing tended to be crystal clear but single-channel. He foresaw that these respective strengths and weaknesses would dovetail to form a new kind of artificial intelligence. He went on to facilitate the development of time-sharing computer systems, toward the goal of computers that could respond to multiple simultaneous inputs and requests.[23] Licklider's essay was widely referenced, and served as an update to the cybernetic philosophical horizon toward which the combined disciplines called AI could push.

At the same time, computer scientists and electronic engineers working on the development of new kinds of circuits sought to frame their work within the exciting frontier of artificial intelligence. The *neuristor* and the *memistor* were experimental circuits developed to explore computer memory using biomorphic modeling. The heavy use of biophilic language to describe innovations in computing did not sit well with everyone. In 1964 an engineer who had worked on an electrochemical "neuron" wrote a letter to the *Proceedings of the IRE* complaining that a particular concept relating to electronic circuits did not possess enough characteristics in common with nervous systems to be appropriately termed a neuristor.

Notwithstanding the general flight from figurative representation in modernism, the human figure was an essential visual forum for artists engaged in expressing the cultural signification of electronics. *Fig. 9.16* is a postmodern collage using contrasting illustrative and scientific graphic elements to convey a sense of posthumanity; it is also a striking deployment

*(opposite) Fig. 9.16:* THE COVER OF A SPECIAL ISSUE OF *PROCEEDINGS OF THE IRE* ON BIOMEDICAL ELECTRONICS, JUST PRIOR TO THE COINING OF THE TERM "BIONICS" (1959).

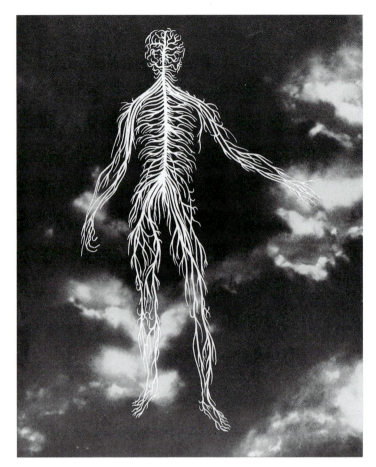

*Fig. 9.17:*
THE MARTIN COMPANY
(*ELECTRONICS*, 1958).

of the collage technique toward suggesting permeability between human and machine. Less frequently seen in commercial art than other modernist artistic strategies, collage interrupts the continuity of preexisting elements by combining them to create new visual ideas. Compare this image with *fig. i.6*, "Electronics: Techniques for a New World 1: A *Fortune* Collage," for a contrast between the collage of 1943, made to introduce new technologies, and the collage of 1959 that takes those technologies in a very personal direction.

Laboratory investigations into bionics clustered throughout the 1960s around developing sensory extensions, and around developing the capacity of machines to mimic human behavior—the robotic impulse. The most accessible cluster, the projects with obvious relevance to the general public, has always been medical applications of electronics. When NASA announced the Mercury 7 astronaut corps in 1959, the men became both heroic figures and laboratory test subjects. A significant corner of space electronics, as articulated in the *Proc IRE* roster of responsibilities for that field that we saw in chapter 8, involved developing devices to monitor and maintain the human body in an orbital and interplanetary environment.

In 1960 two scientists evaluated the implications for bionic space electronics in a way that few people had anticipated. In a thesis that appears extremely prescient from a twenty-first-century perspective, Manfred Clynes and Nathan Kline explained that in the future it might be smarter to reengineer the human body to withstand the punishing conditions of space, as opposed to trying to build all the necessary protections into spacecraft. In a paper titled "Cyborgs and Space," the scientists proposed integrating respiratory technologies into the human body that would spare us the hard work of breathing in outer space.[24] These and other similar suggestions reversed the traditional telescopic view on space electronics—that its job was to protect and study the astronaut. Its connection to the political climate for science today can be seen in the generous funding for genetic modification that outstrips research into human spaceflight. The paper by Clynes and Kline suggests one path to resolving this current tension.

**LANGUAGE CHANGES**    The electronics and medical communities tried out the term "bionics" at first for only a few years. Like the name of the magazine *Nucleonics,* dedicated to the civil nuclear industry, the word infused a new technological sphere with a dose of logo-futurism. The case of these terms can tell us something about language change and the ways it can unfold. "Electronics," coined at the dawn of the twentieth century, is so durable and ubiquitous that it is now invisible. The term "avionics" for flight electronics was introduced in the 1940s and became quickly established, if never quite so universal as "electronics" (naturally, as its meaning is far more specialized). Its analogs "bionics" and "nucleonics," on the other hand, each took different paths: "bionics" migrated through science fiction, in the process gaining enhanced associations with futurism, and is used today (as ever) to refer to the vanguard of biomedical engineering technologies. "Nucleonics" migrated in the opposite direction, away from the public eye, deep into the civil nuclear power industry, and is now found in inter-industry communications but is not widely recognized by the general public. In addition to the futuristic-sounding "bionics," the field that coalesced in the 1960s also called itself variously "biomedical engineering" and "bioelectronics," and its spokespeople enthused then that "theoretically, it should be possible to replace any malfunctioning part of the human body with its electronic counterpart,"[25] inspiring the "bionic man" of science fiction.

**IN CLOSING**    The future of human journeys beyond Earth's orbit may include live human spaceflight missions to asteroids and to Mars, or back to the moon. Whether or not such journeys take place is at this point more a cultural question than one of technology. Deep space telescopes that scan the electromagnetic spectrum for galactic phenomena are providing a spectacular return on our investment into astronomical electronics. At the same time, we have the technology to send human beings off-Earth again if we decide to. What's keeping us hewing to Earth orbit is a lack of resource coordination, not a lack of technology. But no matter who goes beyond Earth orbit and when, those astronauts will likely have dozens, perhaps millions, of ride-along colleagues.

Enlarged view of a memory core plane used in electronic computers.

## "Brain Cells" Of A Billion-Fact Mind
### —FROM RCA RESEARCH IN MAGNETICS

*Fig. 9.18:*
RCA. THE AD COPY PROMOTES
RCA'S WORK IN MAGNETIC
CORE MEMORY FOR
COMPUTERS. IT READS: "MADE
POSSIBLE BY ABACUS-LIKE
ARRAYS OF RCA 'MEMORY
CORES,' THE 'BRAIN CELLS' OF
AN ELECTRONIC MIND
. . ." ALSO NOTEWORTHY AS
A DEPICTION OF A WOMAN
COMPUTER OPERATOR
(*SCIENTIFIC AMERICAN*, 1964).

These colleagues will be safe from damaging radiation and the sickening effects of long-term exposure to low gravity that plagues interplanetary travel. Their perches will instead be comfortable. For these fellow travelers will participate in spaceflight through telepresence, facilitated by an Earth-based electronic environment that will project them into the spacecraft, perhaps even letting them handle tools for exploration. Today the human crews of our robotic Mars rovers are using telepresence to control their craft. Perhaps in the future there will be public access rovers, like public access television stations, in which any person or group of people in the world may apply to participate in scientific exploration as drivers, or even propose a research project for the rover to conduct.

The emerging future of human-electronic integration is progressing in a larger sense toward transhumanism, the state in which the combinatory result from our species-wide encounter with advanced technology is resulting in a new and hopefully improved state of being. In many instances, transhumanism is already here (it is just unevenly distributed).[26] My grandfather lived a decade longer with a pacemaker than he would have otherwise, while my pocket device exerts an indirect influence on the rhythm of my day. My awareness that the device may sound an alarm usually provokes my internal, organic chronometer to keep excellent time: I typically turn off electronic reminders right before they sound. Which of us, my grandfather or me, is not bionic?

On the other side of the coin, we have developed ever more sophisticated forms of robotic and cybernetic warfare. The fiscal mother's milk of Cold War military contracts contradicted Norbert Wiener's faith that cybernetics would yield fundamentally peaceful, democratic, self-organizing systems. The exciting transhumanist technological investigations of that era are the forerunners of our own personal tech, of course, but also of weapons that are as small and nimble as houseflies, and cyborg soldiers of all genders. Far more subtle than the bomb, the development of roboticized and networked warfare nevertheless ranks near it among the most urgent social problems to be solved. And it turned out that the epic war of surveillance, the Cold War, was only a prologue to the culture of total surveillance that we live in today.

Even as these changes have developed in the past half-century, the component parts of new devices have become ever more impressive improvements over previous models, rather than game-changers. The new technologies that have truly recast our frame around what technology is, and can do, like nanotech and the potential of quark computing, are so invisible and still so disengaged from everyday life as to be ill-suited as subjects of visual art (though the narrative art of science fiction is hard at work interpreting these technologies for us).

The role of graphic artists to contextualize new technologies changed dramatically in the mid-1960s. Photography and quotidian illustration (as in *fig. 9.18*) combined in those years to eclipse the golden era of hand-drawn graphic design. Photography was "modernism's way around the distortions of subjectivity" and was "embraced by modernism as a technical form of representation."[27] As such, photography was the slow-developing triumph of modernism in commercial representation. It took decades for this technology of representation, with its technical costs, to become as accessible as, then *more* accessible than, drafting-board graphic design. But by the 1960s photography was the dominant medium of representation in printed advertisements. At the same time, as electronics became increasingly integrated with everyday life, the modern moment became postmodern: The network, and the scope of possibilities that it represents, became a disembodied phenomenon, not unlike a physical force field, that ties us to our technology. That network underpins a visually saturated world that defeats any twenty-first-century attempt to compose a coherent graphic tradition.

In terms of electronic devices, graphic artists have done their jobs. The role of the designer has shifted almost wholly into the domains of industrial design, graphic user interfaces, user experience, and data visualization, as well as the design of systems themselves. Art-for-industry has become integrated into the engineering environment; the most popular engineers today are the total engineers, those who can create functionalities that are either invisible or that offer seamless and aesthetically pleasing user experiences. The most powerful art made today in any context is that which responds to a saturated visual field with strategies like collage, appropriation, annotation, and the creative manipulation and navigation of information and ideas.

**TOWARD A PLANETARY MODEL**

The prospect of electronic circuits that can dissolve within our bodies invokes the journey of the atom and the planet across dramatic steps in scale. Nanoscale bioelectronics represent not just a return to the atom as a reference point but a journey beyond the atom—within it, to subatomic particles, particles our bodies can metabolize and absorb. When science and art attempted to navigate, together, the question of whether emerging electronic technologies should be framed as organic phenomena, like the cohesion of the solar system, or as mechanical systems, the overwhelming evidence of mechanization won out in the 20th century. As the creative tension between the atom and the planet expanded to encompass the human body, the symbolic significance of Earth itself has diminished.

The *Earthrise* photograph, taken by the astronauts of Apollo 8 on December 24, 1968, is legendary for helping to launch the modern environmental movement.[28] Astronomical sciences powered by electronics have turned Earth-centered geoscience into a planetary science. Our developing knowledge of Earth, including climate science, benefits enormously from our exploration of Mars and the rest of the solar system. And yet space sciences are less well supported now than the need to understand the planet demands. In the prevailing machine model of electronics, the electronics industry itself developed in a condition of disconnectedness from feedback loops with society and earth science, allowing for exploitation of labor and the environment.[29] Today's ecology of electronics is badly tilted in favor of production and proliferation and against smart de-production—the downscaling and greening of manufacturing.[30]

I have hoped to show many things in this book, but most salient among them is the power of art to create the world as we wish to see it. Technology does this too; it quite often outruns the designs of its inventors.[31] Kevin Kelly's book *What Technology Wants* explores this phenomenon, and shares some of its logic with critics of technology.[32] Today the culture of technology is one of not-being-seen, a strategy that among other things hides the designs of technology in a cloak of invisibility. The companies that build networked systems and their engineers are trying to convince us that those networks are weightless confections, all clouds and sleek biophilic devices. These strategies seek to deny the physicality of electronic landscapes, when in reality the social and ecological problem of electronic waste is profound and global in its scope.[33] Given the roots of automation in self-correcting industrial systems, the inability of electronic industries themselves to incorporate feedback from humanity and the planet into their systems is ironic evidence in favor of external intervention.

Our systems of unmaking need to develop in the same directions as those dissolvable circuit boards. Let the tension between organic and mechanical systems evolve toward a planetary model that takes ecology and society into account, a future that is dense with organic, biodegradable machines, when computers can be grown in public fields and in backyards through seed kits, and when obsolescent models become useful as fertilizer for next year's crop . . .

It could happen. Or allow me at least to propose that such scenarios are more plausible now than they have ever been. In the past decade there has been an intense resurgence of the connection between handcraft and electronics. Around the world, maker spaces, Maker Faires, tinkering studios, and legacy craft fairs are booming, and not only with analog materials. Children can make conductive circuits using cloth and other soft materials to create their own illuminated and robotic toys, and soldering benches are de rigueur at science museums. Mesh networks based on open source systems allow connections between people to flourish where they are underserved by established infrastructure.[33] This is where art meets electronics with an enthusiasm that again moves in the direction of the future: in the realm of crafts, in three-dimensional, all-access, unbounded formats that possess the ability to surprise at every turn. If compostable computers are going to grow, this is where their medium will be found.

# NOTES

## ABBREVIATIONS USED

*Proceedings of the Institute of Radio Engineers: Proc IRE*
*Proceedings of the Institute of Electrical and Electronic Engineers: Proc IEEE*

## INTRODUCTION

1 For a discussion of how the telephone affected the everyday experience of time and space, see Stephen Kern, *The Culture of Time and Space 1880-1918* (Cambridge, MA: Harvard University Press, 2003), 91-92.

2 "Ten Years of Radio Sales," *Electronics*, March 1932, 78; "Statistics of Radio and Sound-Pictures," ibid., 79.

3 A full discussion of this phenomenon is in R. Roger Remington and Lisa Bodenstedt, *American Modernism: Graphic Design, 1920 to 1960* (New Haven: Yale University Press, 2003), in particular the chapter "A New Style: American Design at Mid-Century 1950-1959," which includes references to some large electronics firms and their programs for corporate identity.

4 "Electronics: A Lever on Industry," *Fortune* 28 (November 1943): 133-35, 198, 200, 202, 205.

5 "Electronics: Techniques for a New World 1: A *Fortune* Collage," *Fortune* 28 (November 1943): pull-out poster.

6 Peter Vardo was credited for the *Fortune* collage as "Vardø." Elsewhere, and when referred to in the historical literature, his name appears as "Vardo."

7 Michael Leja, *Reframing Abstract Expressionism: Subjectivity and Painting in the 1940s* (New Haven and London: Yale University Press, 1993), 73-76, 327.

8 Linda Hales, "Object Lesson: *Fortune*'s Bold Forecast," *Architect* 95, no. 11 (Mid-October 2006): 119.

9 "The Electronics Market: A Key Market Study," *Industrial Marketing* 49 (May 1964): 105-58.

10 "Orestes H. Caldwell Dies: Pioneer in Commercial Radio," *New York Times*, August 30, 1967. Digital file accessed October 2011.

## CHAPTER ONE: THE ATOM, THE PLANET, AND THE TUBE

1  One-page promotion for advertising space in the magazine. *Proc IRE* 42 (April 1953): 48a.

2  Lawrence Lessing, *A Man of High Fidelity: Edwin Howard Armstrong* (Philadelphia and New York: Lippincott, 1956), 213-39.

3  GE had to secure cooperation from the FCC, which had to express willingness to establish and license FM broadcast stations before a gamble could be taken on the development of broadcast technology or the manufacturing of receivers. This process was far from a given, as RCA engaged in several years of litigation and political lobbying to prevent the FCC from opening FM frequencies to regulated access.

4  H. P. Thomas and R. H. Williamson, "A Commercial 50-kilowatt Frequency-Modulation Broadcast Transmitting Station," *Proc. IRE* 29 (October 1941): 537-45.

5  K. C. Dewalt, "Three New Ultra-High-Frequency Triodes," *Proc IRE* 29 (September 1941): 475-80.

6  C. T. Coiner, in *Herbert Bayer: Exhibition of Paintings, Posters, Montages, Advertising Design,* catalog to accompany March 1943 exhibit at North Texas State Teachers College, Denton, TX. Herbert Bayer Papers, Box 1, Denver Art Museum archives.

7  From a notecard clipped to the cover of a copy of the booklet "Electronics: A New Science for a New World," in Owen D. Young Papers, Box 22, Folder 2-18-3A, October 1942–October 1948, GE archives at Schenectady Museum and Planetarium.

8  Peder Anker, "Graphic Language: Herbert Bayer's Environmental Design," *Environmental History* 12 (April 2007): 255-6. See also the illustration of Bayer's globe-shaped installation at the Bauhaus School in 1928, in John Willett, *Art and Politics in the Weimar Period* (New York: Pantheon, 1978), 138.

9  Bayer's work for General Electric is consistent with the theory of being and spatiality expounded by the German philosopher Peter Sloterdijk. Sloterdijk's recent three-volume opus on human spatiality bases his interpretation in a theory of spheres: "The theory of spheres is a morphological tool that allows us to grasp the exodus of the human being from the primitive symbiosis to world-historical action in empires and global systems. . . . It reconstructs the phenomenon of advanced civilization as the novel of sphere transference from the intimate minimum, the dual bubble [of womb and mother], to the imperial maximum, which one should imagine as a

monadic round cosmos. If the exclusivity of the bubble is a lyric motif, the inclusivity of the orb is an epic one." Peter Sloterdijk, *Spheres Volume I: Bubbles / Microsphereology,* trans. Wieland Hoban (Los Angeles: Semiotext(e), 2011), 67.

10 Anker, "Graphic Language," 254-78.

11 Richard A. Bartlett, *The World of Ham Radio, 1901-1950* (Jefferson, NC, and London: McFarland, 2007), 156.

12 Western Electric, *Aviation Communication* (1929), 5.

13 D. K. Martin, "Laying a Foundation for Aircraft Communication," *Bell Laboratories Record* 7 (1929): 1-5.

14 Harold G. Bowen, *The Edison Effect* (West Orange, NJ: Thomas Alva Edison Foundation, 1951), 10-11.

15 Lee de Forest, "Recent Developments in the Work of the Federal Telegraph Company," *Proc. IRE* 1 (January 1913): 37-57.

16 Merle Duston, *Electronics for Beginners* (Detroit: Hoover Industries, 1946), 4.

17 David Sarnoff, *Opportunities in Radio and Electronics for Returning Service Men* (New York: Radio Corporation of America, 1945).

18 IBM's partnership with the artists Charles and Ray Eames, for example, yielded a number of creative fruits in the 1960s that shaped broad public engagement with science, reaching far outside the corporate environment. One of these was "Mathematica: A World of Numbers . . . And Beyond" (1961), an interactive exhibit created for the California Museum of Science and Industry and sponsored by IBM. Another was the film *Powers of Ten* (1967), which explored the expanded sense of scale opened up by scientific inquiry. For a full accounting of IBM's role as a facilitator of design, including art, architecture, and industrial design, see John Harwood, *The Interface: IBM and the Transformation of Corporate Design, 1945-1976* (Minneapolis: University of Minnesota Press, 2011).

19 For a more detailed explanation of this early history, see Michael R. Williams, "A Preview of Things to Come: Some Remarks on the First Generation of Computers," in Raúl Rojas and Ulf Hashagen, eds., *The First Computers: History and Architecture* (Cambridge, MA: MIT Press, 2000), 1-17.

20 For a history of the business machine companies that developed into computer titans in the mid-twentieth century, see James W. Cortada, *Before the Computer: IBM, NCR, Burroughs and Remington Rand and the Industry They Created, 1865-1956* (Princeton, NJ: Princeton University Press, 2003).

21 "Copy Chasers," *Industrial Marketing* 44 (September 1959): 229.

## CHAPTER TWO: TUBES THAT SEE: CATHODE-RAY TUBES

1 Quoted in Richard W. Hubbell, *4000 Years of Television: The Story of Seeing at a Distance* (New York: Putnam, 1942), xi.

2 The oscilloclast was a device that applied variable electric current to the human body for therapeutic purposes. The oscilloscope was applied to monitor that current in a medical context. In Roger Manley and Tom Patterson, *Farfetched: Mad Science, Fringe Architecture and Visionary Engineering* (Raleigh, NC: Gregg Museum of Art and Design, 2013), 10.

3 This observation originated with Kevin Hamilton and Ned O'Gorman at the University of Illinois. Personal conversations, February 2011.

4 Edward L. Safford, Jr., *Modern RADAR: Theory, Operation, and Maintenance* (Blue Ridge Summit, PA: Tab Books, 1981).

5 G. F. J. Garlick, S. T. Henderson, and R. Puleston, "Cathode-Ray-Tube Screens for Radar," *Journal of the Institution of Electrical Engineers* 93 (May 1946): 815-21.

6 David P. D. Munns, *A Single Sky: How an International Community Forged the Science of Radio Astronomy* (Cambridge, MA, and London: MIT Press, 2013).

7 "Television: A Survey of Present-Day Systems," *Electronics*, October 1934, 300-305.

8 George Everson, *The Story of Television: The Life of Philo T. Farnsworth* (New York: Norton, 1949). Also see Albert Abramson, *Zworykin: Pioneer of Television* (Urbana, IL, and Chicago: University of Illinois Press, 1995), and Hubbell, *4000 Years of Television*.

9 "Television must be available to all the people at once. Otherwise the radio business will be killed in those cities anticipating the coming of a television service already in use elsewhere. Radio is not in a position to receive such a blow." "Television: A Survey of Present-Day Systems," 304.

10 David Weinstein, *The Forgotten Network: DuMont and the Birth of American Television* (Philadelphia: Temple University Press, 2006).

11 Ned O'Gorman and Kevin Hamilton, "At the Interface: The Loaded Rhetorical Gestures of Nuclear Legitimacy and Illegitimacy," *Communication and Critical/Cultural Studies* 8 (2011): 41-66.

12 Solomon R. Guggenheim Museum, "Conservation: Time-Based Media," http://www.guggenheim.org/new-york/collections/conservation/time-based-media (accessed July 2014).

13 *Electronics*, March 1932, 101.

14 *Wavemakers / Les Chant des Ondes*, directed by Caroline Martel (Artifact Productions, 2013), film. Viewed at the Exploratorium, San Francisco, February 20, 2014. Also personal conversations with Caroline Martel, February 19-22, 2014.

15 *Wavemakers / Les Chant des Ondes,* synopsis, http://artifactproductions .ca/lechantdesondes/wavemakers-synopsis/?lang=en (accessed February 2014).

16 "The New Music of Electronic Oscillations," *Electronics*, March 1932, 86-87, 114.

17 Frank H. Rockett, "The Electron Art," *Electronics*, January 1949, 130.

18 "Ultrafax," ibid., 77-79.

19 "Oscillograms As a Logical Form of Written Speech," *Electronics*, January 1932, 24.

20 J. P. Eckert, Jr., H. Lukov, and G. Smoliar, "A Dynamically Regenerated Electrostatic Storage System," *Proc IEEE* 38 (May 1950): 498-510.

21 J. P. Eckert, Jr., "A Survey of Digital Computer Memory Systems," *Proc IEEE* 41 (October 1953): 1393-406.

22 M. L. Dertouzos, "CIRCAL: On-Line Circuit Design," *Proc IEEE* 55 (May 1967): 637-53.

23 Jeremy Bentham, *Panopticon; Or, the Inspection-house* (New York: General Books, 2010).

## CHAPTER THREE: COLD ROCK, WARM LIFE: CRYSTALS

1 "Radio Uses of Piezo-Electric Crystals," *Journal of the AIEE* 44 (December 1925): 1290.

2 This paragraph in particular, and to a lesser extent the general historical background of this chapter, relies on Michael Riordan and Lillian Hoddeson, *Crystal Fire: The Invention of the Transistor and the Birth of the Information Age* (New York: Norton, 1997).

3 Hendrik B. G. Casimir, "Maxwell, Hertz, and Lorentz," in Carl F. J. Overhage, ed., *The Age of Electronics* (New York: McGraw-Hill, 1962).

4 *Industrial Marketing* 44, no. 4 (April 1959): 9-11.

5 Edmund Whittaker, *A History of the Theories of Aether and Electricity*, vol. 1, *The Classical Theories* (New York: Harper & Brothers, 1960).

6 Isaac Koga, "Note on the Piezo-electric Quartz Oscillating Crystal Regarded from the Principle of the Similitude," *Proc IRE* 19 (June 1931): 1022-3. Also *Crystals* (Bell Laboratories, 1958), film, Internet Archive, Prelinger Archives collection, https://archive.org/details/Crystals1958 (accessed October 15, 2013).

7 "Radio Uses of Piezo-Electric Crystals," 1290.

8 Walter G. Cady, "Crystals and Electricity," *Scientific American* 181 (December 1949): 46-51.

9 "Piezo-Crystal Phonograph Reproducer," *Electronics*, June 1934, 198.

10 *Crystals Go to War* (Reeves Sound Laboratories, 1943), film, Internet Archive, Prelinger Archives collection, https://archive.org/details/6101_Crystals_Go_to_War_01_20_16_21 (accessed August 15, 2013).

11 "Manufacture of Quartz Crystals," *Electronic Industries* 2 (May 1943): 58-62.

12 *Crystals Go to War.*

13 *Crystals.*

14 Alfred L. Loomis and W. A. Marrison, "Modern Developments in Precision Timekeepers," *Journal of the AIEE* 32 (August 1932): 542-49.

15 Ignacio Chapela, "Liberation Biology and Mutual Aid," paper delivered at Anthropology and Social Change second annual conference, California Institute of Integral Studies, March 21, 2014.

16 Walter Murch, the painter, was the father of the filmmaker Walter Murch. Since the advent of his son's career, Murch Senior has been referred to as Walter Tandy Murch. I am indebted to the Murch family, in particular Beatrice Murch and Walter Murch, for their helpfulness with regards to my inquiries about their artistic forebear.

17 Cady, "Crystals and Electricity."

18 A. C. Walker and G. T. Kohman, "Growing Crystals of Ethylene Diamine Tartrate," *Transactions of the AIEE* 67 (January 1948): 565-70.

19 Daniel Robbins, "Walter Murch," in *Walter Murch: A Retrospective Exhibition* (Rhode Island School of Design Museum of Art, 1966), unnumbered pp. 1-41.

20 Ernest W. Watson, "Walter Murch—Painter," *American Artist* 19, no. 8 (October 1955): 22-27, 62-63.

21 Cyril Stanley Smith, *A History of Metallography* (Chicago: University of Chicago Press, 1960).

22 Sophie Forgan, "Festivals of Science and the Two Cultures: Science, Design and Display in the Festival of Britain, 1951," *British Journal for the History of Science* 31 (June 1998): 217-40, http://journals.cambridge.org/article_S0007087498003264 (accessed August 2013).

23 Cyril Stanley Smith, "Structure, Substructure, Superstructure," in György Kepes, ed., *Structure in Art and Science* (New York: George Braziller, 1965), 29-41.

24 György Kepes, *Language of Vision* (Chicago: Paul Theobald, 1951), 28-29.

25 Albany Institute of History and Art, *Art in Science* (1965): exhibition catalog.

26 Robert M. Coates, "The Art Galleries," *New Yorker*, October 16, 1965.

27 Robert Hughes, "Art," *Time*, October 11, 1971, 86-87.

## CHAPTER FOUR: TRANSISTORS AND CIRCUIT SYMBOLS

1 J. A. Morton and W. J. Pietenpol, "The Technological Impact of Transistors," *Proceedings of the IRE* 46 (June 1958): 955-59.

2 See ibid. for more detail about this chronology.

3 J. Bardeen and W. H. Brattain, "The Transistor, a Semi-Conductor Triode," *Physical Review* 74, no. 2 (1948): 230.

4 J. Bardeen and W. H. Brattain, "Physical Principles Involved in Transistor Action," *Physical Review* 75, no. 8 (1949): 1208.

5 "Transistorized Auto Radio Developed by RCA Scientists," *Technician and Circuit Digest* 61, no. 5 (May 1955): 16.

6 The area now known as Silicon Valley received a boost through the arrival of Eastern Bell Laboratories alumni who set up shop there in the mid-1950s. The region had been a leader in electrical and electronic engineering for decades, a result of the suitability of the marshy landscape to radio transmission experimentation, the long-standing presence of a Pacific-oriented U.S. military establishment, the presence of Stanford University, and other factors. Lee de Forest had developed the first audion tube amplifier in Palo Alto in 1906, and in 1955 more than fifty electronics laboratories and firms were established in Silicon Valley, including familiar names such as Hewlett- Packard, IBM, Sylvania, and Varian Associates. It was a historical coincidence that William Shockley had been born in Palo Alto and returned there following his tenure at Bell Laboratories, a move that resulted in an enormous infusion of new scientific research energy. See "1908-1955: Radio, Radar, Television Electronics Era Began in Palo Alto," *The Tall Tree* 1, no. 7 (October 1955): 3-23; Arthur L. Norberg, "The Origins of the Electronics Industry on the Pacific Coast," *Proc IEEE* 64 (September 1976): 1314-22.

7 R. M. Ryder and R. J. Kircher, "Some Circuit Aspects of the Transistor," *Bell System Technical Journal* 28 (July 1949): 367-400.

8 The story of the graphic symbol for the transistor is more complex than this particular narrative permits. Early in the 1950s transistors were assembled with different kinds of internal structures, and the symbols for these different types of transistors varied based on their structural differences. Later, with the development of integrated circuits and the leap

from mesa to planar transistor, the symbol changed again and lost its encompassing circle. As transistors were assembled into networks of dozens and hundreds within a circuit board, the symbol was correspondingly simplified.

9  "IRE Standards on Reference Designations for Electrical and Electronic Equipment, 1957," *Proc IRE* 45 (November 1957): 1493-501; Howard L. Cook, "Symbols for Electricity and Electronics," *IEEE Transactions on Engineering Writing and Speech* 8 (December 1965): 60-6; "A Graphic Language for Those Concerned with Eectrical Diagrams," *Journal of the Society of Motion Picture and Television Engineers* 63 (September 1954): 107.

10  Philip B. Meggs, *A History of Graphic Design* (New York: Wiley, 1998), 213-23.

11  Walter Crane, *Line and Form* (London: G. Bell, 1900) https://archive.org/details/lineform00cranuoft (accessed September 2012).

12  James R. Harris, "The Earliest Solid-State Digital Computers," *IEEE Annals of the History of Computing* 21, no. 4 (1999): 49-54; Morton and Pietenpol, "The Technological Impact of Transistors."

13  Donald G. Fink, "Crosstalk" (column), *Electronics*, September 1948, 67.

## CHAPTER FIVE: CIRCUIT BOARDS AND THE MATRIX

1  Joseph Whitehill, *The Angers of Spring* (Boston and Toronto: Little, Brown, 1959), 212.

2  "Radio Robot Squirts Out Three a Minute," *Popular Science*, April 1948, 160-61.

3  J. A. Sargrove, "New Methods of Radio Production," *Journal of the British Institution of Radio Engineers* 7 (January-February 1947): 2-33.

4  "Printing Electronic Circuits," *Electrical Engineering* 66 (April 1947): 351; S. F. Danko and R. A. Gerhold, "Printed Circuitry for Transistors," *Proc. IRE* 40 (November 1952): 1524-28.

5  N. A. Skow, "Laminates for Printed Circuits," *Electrical Engineering* 74 (December 1955): 1092-93.

6  Robert L. Swiggett, "Printed Circuits on Foil-Clad Plastics," *Modern Plastics* 28 (August 1951): 99-107, 110-11.

7  Lewis Mumford, "From Handicraft to Machine Art," in *Art and Technics* (New York: Columbia University Press, 1952), 63.

8  W. H. Hannahs and J. W. Eng, "Production Control of Printed Resistors," *Electronics*, October 1952, 106-10.

9  "A Review of Procedures Found Useful at Photocircuits Corporation for Making Printed Circuit Masters," *Graphic Science* 1 (October 1959): 16-22.

10 Danko and Gerhold, "Printed Circuitry for Transistors," 1524.

11 Personal communications, 2009.

12 *Bell Telephone Magazine* 63 (Spring 1964): 61. Stand-alone photo feature with caption.

13 "Circuitry that Bends Takes a Sales Upturn," *Business Week*, February 12, 1966, 150–51.

14 Jack S. Kilby, "Invention of the Integrated Circuit," *IEEE Transactions on Electron Devices* ED-23 (July 1976): 648–54.

15 For a complete account of this history, see Riordan and Hoddeson, *Crystal Fire*.

16 *The Fairchild Chronicles*, DVD (Walker Research Associates, 2005). © Stanford University Library, 2005.

17 Guy B. Senese, *Self-Determination and the Social Education of Native Americans* (Westport, CT: Greenwood, 1991), 77–79; David Dworsky, "Industry and the Indian," *New York Times*, July 23, 1967.

18 Lisa Nakamura, "Indigenous circuits," blog post, January 2014, http://www.computerhistory.org/atchm/indigenous-circuits/ (accessed February 2014).

19 "Industry Invades the Reservation," *Business Week*, April 4, 1970: 72.

20 Senese, *Self-Determination and the Social Education of Native Americans*, 79.

21 See Jefferson Cowie, *Capital Moves: RCA's Seventy-Year Quest for Cheap Labor* (Ithaca, NY, and London: Cornell University Press, 1999), and Jennifer Gabrys, *Digital Rubbish: A Natural History of Electronics* (Ann Arbor, MI: University of Michigan Press, 2011).

22 Jack S. Kilby, "Turning Potential Into Realities: The Invention of the Integrated Circuit," Nobel Lecture, December 8, 2000, http://www.nobelprize.org/nobel_prizes/physics/laureates/2000/kilby-lecture.html (accessed June 2012).

23 Yates McKee, "The Public Sensoriums of Pulsa: Cybernetic Abstraction and the Biopolitics of Urban Survival," *Art Journal* 67 (Fall 2008): 46–67; Jim Burns, *Arthropods: New Design Futures* (New York: Praeger, 1972), 138–42.

24 Lucy Lippard, *Six Years: The Dematerialization of the Art Object from 1966 to 1972* (Berkeley, CA: University of California Press, 1997), 81–82.

25 Gordon E. Moore, "Cramming More Components onto Integrated Circuits," *Electronics*, April 19, 1965, 114–17.

26 H. Maeda, A. Matsushita, and M. Takashima, "Woven Wire Memory for

NDRO System," *IEEE Transactions on Electronic Computers* EC-15 (August 1966): 442–51.

27 J. L. Goldberg and R. J. Slaughter, "Braid Construction and Attenuation of Coaxial Cables at Microwave Frequencies," *Proc IEEE* 113 (June 1966): 957–62.

28 Personal communication, December 13, 2013.

## CHAPTER SIX: AUTOMATIC AND DIGITAL: THE EMERGENCE OF COMPUTING

1 Robertson T. Barrett, "A Century of Electrical Communication," *Bell Telephone Magazine* 23 (Spring 1944): 43–51.

2 The historian Michael Sean Mahoney has pointed out that "the computer is not one thing, but many different things"; for more on this perspective on the history of computing, see Mahoney, *Histories of Computing*, edited and with an introduction by Thomas Haigh (Cambridge, MA, and London: Harvard University Press, 2011).

3 *Communications of the Association of Computing Machinery* 1, no. 6 (June 1958): 21.

4 In this chapter I am reverting to the anachronistic term "digitalization" because it refers specifically to changes made at the broadest levels of technologies and systems—the change from an analog system to a digital system. The use of this term, while archaic to contemporary ears, nevertheless steers us clear of confusion with the familiar contemporary discourse of *digitization*, which refers to the translation of stored information from analog formats such as printed books and LP records into digital formats.

5 Scott Schaut, *Robots of Westinghouse, 1924–Today* (Mansfield, OH: Mansfield Memorial Museum, 2006), 21–22.

6 J. Presper Eckert, Jr., "A Survey of Digital Computer Memory Systems," *Proceedings of the IRE* 41 (October 1953): 1393–1406.

7 The earliest analog computers were human beings who solved mathematical problems in longhand. The slide rule is the most tangible and familiar example of an analog computer. Many other devices abound. See David Alan Grier, *When Computers Were Human* (Princeton, NJ: Princeton University Press, 2007). For an early circle-shaped analog vector computer, see M. P. Weinbach, "Abridgment of Vector Calculating Device," *Journal of the AIEE* 47 (May 1928): 336–40.

8 In 1960 a mathematician named H. Kauner conducted experiments in the conversion of integers using an abacus, published as "A Note on the Use

of the Abacus in Number Conversion," *Communications of the Association of Computing Machinery* 3 (March 1960): 167.

9 Donald Fink, "Crosstalk" (column), *Electronics*, October 1948, 67.

10 F. P. Brooks, Jr., G. A. Blaauw, and W. Buchholz, "Processing Data in Bits and Pieces," *IEEE Transactions on Electronic Computers* 8 (June 1959): 118-24.

11 The introduction of automation in early nineteenth-century England was commensurate with widespread displacement of millions of people from their homes through a government policy of mass evictions. The two factors compounded to yield a devastating enclosure of both land and work against the skilled hands of erstwhile craft- and tradespeople. Against this wholesale disenfranchisement, the practice of machine-breaking as political revolt by the revolutionary Luddites became associated with a stereotype of a person holding a reflexive, uninformed bias against technology. The historical reality is far more complex, as is often the case. In the twenty-first century that pattern has been constructively problematized by the increasing use of social media as a tool of political activism, but it remains entrenched in the realm of manufacturing—the labor used to manufacture electronic devices, and the harmful impact that much electronics manufacturing and waste processing has on ecological systems.

12 Frederick J. Rex, Jr., "Herman Hollerith, the First 'Statistical Engineer,'" *Computers and Automation* 10 (August 1961): 10-13.

13 The 1890 census was used, for example, by the historian Frederick Jackson Turner in his landmark essay "The Significance of the Frontier in American History." The census is also well-remembered in the field of archives, as the enormous scope and volume of its materials defied the limits of the archival practices of the day. In January 1921, much of the original data collected for the 1890 census was lost in a fire. See Robert L. Dorman, "The creation and destruction of the 1890 federal census," *American Archivist* 71 (Fall/Winter 2008): 350-83.

14 Philip J. Klass, "Punch Card Control Gives Auto-Assembler Flexibility," *Aviation Week* 62 (May 2, 1955): 12-14.

15 Recall the rallying cry "I am not a number! I am a free man!" shouted by the main character of the BBC TV show *The Prisoner* (1966-68) during the opening sequence of every episode. One of that show's many hooks was the enduring mystery about who, or what, held the Prisoner prisoner. His isolation and helplessness were the products of a system so vast, yet so invisible, that it was as impermeable as it was abstract—an apt analogy for automated information systems.

16 There has been considerable scholarly attention paid to the many uses of punched cards. See Steven Lubar, "Do Not Fold, Spindle, or Mutilate: A Cultural History of the Punch Card," *Journal of American Culture* 15, no. 4 (Winter 1992): 43–55, http://www.design.osu.edu/Carlson/history/PDFs/lubar-hollerith.pdf (accessed December 2013). Notably, IBM punched cards were used by Germany during World War II to organize concentration camps. See Edwin Black, *IBM and the Holocaust: The Strategic Alliance between Nazi Germany and America's Most Powerful Corporation* (New York: Crown, 2001).

17 Conlon Nancarrow, *Conlon Nancarrow: The Original 1750 Arch Recordings*, compact disc, Other Minds, 2008. The CD collects four preexisting LPs: *Complete Studies for Player Piano*, vols. 1, 2, 3, and 4 (originally published by 1750 Arch, Inc., 1977, 1979, 1981, and 1984, respectively).

18 Edmund C. Berkeley, *Giant Brains, or, Machines That Think* (New York: Wiley, 1949), 119.

19 Frank J. Romano, *Condensed Handbook of Composition Input* (Salem, NH: Graphic Arts Marketing Associates in conjunction with National Composition Association, n.d. [ c. 1975]). This helpful textbook was produced to give keypunch operators, typesetters, and computer data entry workers all the information they would need to read, code to, and translate between any punched card and paper tape coding system they might encounter in their professional lives. I am especially indebted here to Rick Prelinger, a professional phototypesetter between 1977 and 1982, who originally owned and used this book, for his help in understanding punched card and paper tape coding.

20 See, for example, Robert V. Lewis, "Top Management Participation: Key to a Successful Electronic Data Processing Program," *Data Processing* 3 (October 1961): 34–35. The language across trade magazines including *Datamation*, *Data Processing*, and *Electronics* is consistent in referring to mainframe computers as "data-processing units," or similar, until the mid-1960s.

21 Claude R. Kagan, "Computer Coding, Multiplexing, and Distribution of Telegraph Signals," *IEEE Transactions on Communication Technology* 14 (February 1966): 31–39.

22 The name "Univac" first referred, in the 1940s, to the pioneering computing machine that bore the full name "Universal Vacuum Tube Computer," invented by J. Presper Eckert and John Mauchly, aided by a team of women programmers. By 1950 the name referred to the company formed by the computer's two principal inventors, and was sometimes written in all caps

as if it were an acronym even though it was not. In 1950 the company was acquired by Remington Rand, which then referred to itself as Remington Rand-UNIVAC (see *fig. 1.10*, for example).

23 This information appears in many places, but I relied for rough navigation through this territory on the account of input/output technologies that appears in Andrew Bluemle's *Automation* (Cleveland and New York: World Publishing Co., 1962).

24 "Magnetic Tape to the Rescue," *Economist*, November 30, 2013, 3-4.

25 "New Computers Use More Transistors," *Aviation Week*, April 18, 1955, 87.

26 George Steele and Paul Kircher, *The Crisis We Face: Automation and the Cold War* (New York: McGraw-Hill, 1960).

27 Alvin Toffler, "The Future as a Way of Life," *Horizon* 7, no. 3 (Summer 1965): 108-16.

## CHAPTER SEVEN: VISIBLE LANGUAGE

1 George Reitweisner, "An ENIAC determination of $\pi$ and $e$ to more than 2000 decimal places," *Mathematical Computation* 4 (January 1950): 11-15. The computational project was conceptualized by von Neumann, but it was executed using punched cards by a team of four people who worked through the weekend alternating eight-hour shifts to keep the machine running continuously. Those people were ENIAC staff members Clyde V. Hauff, W. Barkley Fritz, and (Miss) Homé S. McAllister, in addition to Reitweisner.

2 P. Henrici, *Elements of Numerical Analysis* (New York: Wiley, 1964). See also M. L. Dertouzos, "CIRCAL: On-Line Circuit Design," *Proc IEEE* 55 (May 1967): 637-54.

3 Howard Eves, "The Latest About $\pi$," *Mathematics Teacher* 55 (February 1962): 129-30.

4 Philip J. Davis, *The Lore of Large Numbers* (Washington, DC: Mathematical Association of America, 1961), 71. Also, D. Shanks and J. W. Wrench, Jr., "Calculation of $\pi$ to 100,000 Decimals," *Mathematics of Computation* 16 (January 1962): 76-99.

5 Emerson W. Pugh, Lyle R. Johnson, and John H. Palmer, *IBM's 360 and Early 370 Systems* (Cambridge, MA: MIT Press, 1991), 36. Also "News and Notices," *Communications of the Association for Computing Machinery* 2 (February 1959): 34.

6 Stephen Eskilson, *Graphic Design: A New History* (New Haven: Yale University Press, 2007), 147-9, 158-67.

7 See, for example, Georgy and Vladimir Stenberg's undated film poster, c.

1930s, reproduced as fig. 17.30 in Philip B. Meggs, *A History of Graphic Design*, 3rd ed. (New York: Wiley, 1998), 270.

8  Emmett Williams, ed., *Anthology of Concrete Poetry* (1967; reprinted New York: Primary Information, 2013), iv.

9  "Company's growth feeds on unified advertising," *Industrial Marketing* 49, no. 6 (June 1964): 97-99.

10  Stanley P. Frankel, "Logical Design of a Simple General Purpose Computer," *IRE Transactions on Electronic Computers* EC-6 (March 1957): 5-14, and Edward Lorenz, "3 Approaches to Atmospheric Predictability," *Bulletin of the American Meteorological Society* 50 (May 1969): 345-50.

11  Frank Leary, "Computers Today—Part 1: The Industry Today," *Electronics*, April 28, 1961: 63-72. Also, "Part 2: The Technology Today," ibid.: 73-94.

12  Among other sources, this history is recounted in a personal manner by Tom J. Sawyer at http://tjsawyer.com/B205Home.htm (accessed August 2013).

13  "Discussion: The Burroughs B 5000 in Retrospect," *Annals of the History of Computing* 9 (Winter 1987): 37-92.

14  Nathan Ensmenger, *The Computer Boys Take Over: Computers, Programmers, and the Politics of Technical Expertise* (Cambridge, MA: MIT Press, 2010).

15  E. E. David, Jr., and Oliver G. Selfridge, "Eyes and Ears for Computers," *Proc IRE* 50 (May 1962): 1093-101.

16  A history of software development is beyond the scope of this book. For more information, begin with Raúl Rojas and Ulf Hashagen, eds., *The First Computers: History and Architectures* (Cambridge, MA: MIT Press, 2000); Mahoney, *Histories of Computing*; Paul E. Ceruzzi, *A History of Modern Computing*, 2nd ed. (Cambridge, MA, and London: MIT Press, 2003); Ensmenger, *The Computer Boys Take Over*; and Kurt Beyer, *Grace Hopper and the Invention of the Information Age* (Cambridge, MA: MIT Press, 2009).

17  Beyer, *Grace Hopper*.

18  Joseph R. de Paris, "Aids for IBM 1400 Series," *Data Processing* 3 (August 1961): 22.

19  In the course of research for this book, a review of the publications *Computers and Automation* and *Communications of the Association of Computing Machinery* found women participating at high levels, and in solid numbers, throughout the 1950s and into the 1960s. To list just three examples from 1961, five papers on programming at the sixteenth annual ACM conference in July 1961 were presented or copresented by women (*CACM*, July 1961);

the Russian American programmer Ariadne Lukjanow, who founded her own company, Machine Translation, Inc., began to offer Russian-to-English translation services to clients in the Washington, DC, area (*CACM*, August 1961); and the naval programmer Dr. Elizabeth Cuthill was honored "by the nuclear reactor design people when they named a code she had helped to develop the Cuthill Code" (*CACM*, September 1961). These anecdotes are offered as evidence pointing to the need for further investigation, and disputing the common impression that programming was fully masculinized during the 1950s.

20 See, for example, the Litton Industries advertisement in *Computers and Automation* (December 1961), and the full-page Philco advertisement in the January 1962 issue.

21 *Computers and Automation* was published by the professional computer systems consultant Edmund C. Berkeley, author of *Giant Brains*. It therefore lacked the status of the industry-wide professional trade journal *Datamation* (founded in 1957), but it published many recruitment advertisements directly targeted at programmers.

22 The term "artificial intelligence" is attributed to Marvin Minsky, cofounder (with John McCarthy) of the AI Laboratory at MIT. As reported in the *Communications of the Association for Computing Machinery*, Minsky coined the term at the November 1958 National Physical Laboratory Symposium on the Mechanization of Thought Processes, "to replace the more controversial term 'thinking.'" *Communications of the Association of Computing Machinery* 2 (April 1959): 29.

23 An unsubstantiated report of an electronic reading device appears as the small photo feature "Phototube 'Translates' Books for Blind Readers," *Electronics*, September 1935, 44.

24 Oliver G. Selfridge, "Pattern Recognition and Modern Computers," *1955 Proceedings of the Western Joint Computer Conference*, 91–93.

25 Steven H. Unger, "Pattern Detection and Recognition," *Proc IRE* 47 (October 1959): 1737–52.

26 *Computer Speech*, 45rpm record (Bell Laboratories, 1963).

27 "Discussion: The Burroughs B 5000 in Retrospect." This entire issue of *Annals of the History of Computing* is dedicated to articles (mostly reprinted from historical issues of *Datamation*) on the B 5000. The article is a transcript of a daylong twenty-fifth-anniversary meeting.

28 This advertisement appeared in *Scientific American*, *Electronics*, and *Data Processing*, and perhaps elsewhere.

29 In this context, the association with film and theater extends to a particular

association with the tradition of twentieth-century industrial filmmaking, in which many of the large gestures of dramatic cinema were reinvented to serve the development of corporate identity. See Hanna Rose Shell, "Films in the Archive: Hollywood in Detroit," *Technology and Culture* 55 (July 2014): 711-14.

30 While the language change was sometimes understood as an externally assigned term referring to the societal impact of automation, its use was also part of a deliberate attempt by the computer industry to present a more sophisticated front to the world. For example, in 1961 the Eastern/Western Joint Computer Conference renamed itself the American Federation of Information Processing Societies.

31 "New Equipment: Burroughs B 5000," *Data Processing* 3 (March 1961): 39-40.

32 *Lincoln Laboratory* (Cambridge, MA: MIT Office of Publications, 1963), informational booklet.

33 Andrew J. Butrica, *To See the Unseen: A History of Planetary Radar Astronomy,* NASA SP-4218 (Washington, DC: Government Printing Office, 1996), 57-59.

34 "Muriel Cooper, 68, Dies; Noted Graphic Designer," http://web.mit.edu/newsoffice/1994/cooper-0601.html (accessed November 2013).

35 Elizabeth Resnick, "Woman at the Edge of Technology," *Eye* 68, no. 17 (August 2008): 26-28.

36 Dietmar R. Winkler, ed., *Posters, Jacqueline S. Casey: Thirty Years of Design at MIT* (Cambridge, MA: MIT Museum, 1992).

37 "Lincoln Laboratory Advertisements," memorandum from Joseph Mindell to Carl F. Overhage, November 26, 1962. Mindell was assistant director of the laboratory, and Overhage was an advertising agent who coordinated the development of the ads between Jacqueline Casey's artwork and the text and directives supplied by Mindell's office. Lincoln Laboratory Information Services Department, Archives and Documents, Laboratory Staff Recruitment—1960s, Box 1.

38 One of the articles abstracted in this figure is R. F. Soohoo, "Excitation and Boundary Effects in Spin-Wave Resonance," *Journal of Applied Physics* 32 (March 1961): 148S-150S. The others remain unidentified.

39 Paul Ceruzzi, "Electronics Technology and Computer Science, 1940-1975: A Coevolution," *Annals of the History of Computing* 10 (April 1989): 257-75.

40 For example, F. E. Heart and A. A. Mathiasen, "Computer Control of the Haystack Antenna," *Proc IEEE* 54 (December 1966): 1742-51.

41 Neil Mills, from notes regarding "Experiments in Disintegrating Language,"

http://www.ubu.com/sound/konkrete.html (accessed June 2013). Record-ings of Mills reading his number poems can be accessed at the same site. For more on the poetics of numbers, see Sarah Glaz, "The Poetry of Prime Numbers," *Proceedings of Bridges 2011: Mathematics, Music, Art, Architecture, Culture* (Tessellations, 2011), 17-24, http://www.math .uconn.edu/~glaz/My_Articles/ThePoetryofPrimeNumbers.Bridges11.pdf (accessed December 2012).

42 http://www.rand.org/content/dam/rand/pubs/monograph_reports/ MR1418/MR1418.deviates.pdf (accessed December 2012).

43 Kenneth Goldsmith, *Uncreative Writing* (New York: Columbia University Press, 2011), 1.

44 Kenneth Goldsmith, *Traffic* (Editions Eclipse Online, 2006), http://english .utah.edu/eclipse (accessed July 2012).

45 Goldsmith, *Uncreative Writing*, 16.

## CHAPTER EIGHT: THE FURTHEST HORIZON: SPACE ELECTRONICS

1 Simon Ramo, address to the American Institute of Electrical Engineers, February 1958, quoted in *Communications of the Association for Comput-ing Machinery* 1 (May 1958): 21.

2 James M. Grimwood, *Project Mercury: A Chronology*, NASA SP-4001 (Washington, DC: Government Printing Office, 1963), appendix 9, 222-23.

3 William Hines, "500 Contractors Share the Work in Gemini Flight," *Wash-ington Star*, March 21, 1965. NASA History Office archives, NASA contrac-tor roster files, Project Gemini folder.

4 Robert Rosholt, *An Administrative History of NASA, 1958-1963*, NASA SP-4101 (Washington, DC: Government Printing Office, 1966), 58.

5 This was the subject of my previous book, *Another Science Fiction: Adver-tising the Space Race 1957-62* (New York: Blast Books, 2010).

6 The *Ranger 3* spacecraft carried an RCA camera system based on a pho-tomultiplier tube; unfortunately its flight failed because of the impact on its central computer of the preflight heat sterilization routine, causing it to overshoot the moon by 20,000 miles. Its lunar seismograph and gamma ray detector functioned, however, having been assembled sterile and therefore not having been subjected to the same sterilization procedure as the main craft, and returned useful data about the radiologic spectra of the moon. The photomultiplier tube that it carried became the furthest-flying vacuum tube when it passed the moon on its way beyond Earth orbit. See M. A. Van Dilla, E. C. Anderson, A. E. Metzger, and R. L. Schuch, "Lunar

composition by scintillation spectroscopy," *IRE Transactions on Nuclear Science* 9 (1962): 405-12; Mark Williamson, *Spacecraft Technology: The Early Years* (London: Institute of Electrical Engineers, 2006), 211-12.

7 "'Explorer II' takes tubes aloft," *Electronics*, December 1935, 31-4, 38.

8 O. H. Gish and H. G. Booker, "Nonexistence of Continuous Intense Ionization of the Troposphere and Lower Stratosphere," *Proc IRE* 27 (February 1939): 117-25.

9 At least nine scientific and technical papers were produced in the first three years following the flight of *Explorer II*, interpreting the findings of the instruments it carried. See Gish and Booker, "Nonexistence."

10 National Aeronautics and Space Administration, *Exploring Space: Projects Mercury and Apollo . . . of the United States Manned Space Program* (Washington, DC: Government Printing Office, 1961), 18; National Aeronautics and Space Administration, *The X-15 Research Airplane* (Washington, DC: Government Printing Office, 1961).

11 Leston Faneuf, "Electrical Frontiers in Space," *Electrical Engineering* 77 (October 1958): 893-97.

12 Ibid., 896.

13 *The Fairchild Chronicles.*

14 "Navy Vanguard in Orbit," *Electrical Engineering* 77 (May 1958): 475-76.

15 Conrad Hoeppner, "Space Electronics," *Proc IRE* 48 (April 1960): 435-37.

16 James E. Webb, letter to John W. McCormack (Speaker of the House of Representatives), January 31, 1964. In House Committee on Science and Astronautics, *Electronic Research Center: Report of the National Aeronautics and Space Administration*, 1964, Committee Print, v.

17 Prelinger, *Another Science Fiction*, 59-96.

18 Prelinger, *Another Science Fiction*, 59-96.

19 Major A. Johnson, *Progress in Defense and Space: A History of the Aerospace Group of the General Electric Company* (Major A. Johnson, 1993).

20 A. C. Dickieson, "The Telstar Experiment," *Bell System Technical Journal* 4 (July 1963): 739-46.

21 "Project Echo Transmits Telephone Messages Via Satellite," *Bell Laboratories Record*, September 1960.

22 Ibid., 740.

23 Prelinger, *Another Science Fiction*, 34, 46.

24 Philip J. Taubman, *Secret Empire: Eisenhower, the CIA, and the Hidden Story of America's Space Espionage* (New York: Simon and Schuster, 2003), 69.

25 James E. Tomayko, *Computers in Spaceflight*, chapter 7, "The Evolution

of Automated Launch Processing." Available at: http://www.history.nasa .gov/computers/Ch7-2.html (accessed December 2013).

26 Webb to MacCormack, v.

27 Ibid., xi.

28 Andrew J. Butrica, "The Electronics Research Center: NASA's Little-Known Venture into Aerospace Electronics," *The Moon* 3 (2002): 9, http://www .history.nasa.gov/ercaiaa.html (accessed January 2014).

29 Milton J. Minneman, personal communication, September 4, 2008; also Milton J. Minneman, "An Experimental Plasma Propulsion System," *IRE Third National Convention on Military Electronics, Professional Group on Military Electronics* (1959): 167-74.

30 Information taken from the Dawn Mission website, http://dawn.jpl.nasa .gov/mission/ (accessed June 2010).

31 There's a lot of history to this, including the fact that the U.S. had conducted its own atmospheric detonations of nuclear weapons as part of Project Argus in 1958.

32 Charles C. Bates, "Detection and Identification of Nuclear Explosions Underground (Project Vela Uniform)," *Proc IRE* 50 (November 1962): 2201-8; M. G. Gudzin and E. G. Holle, "Seismological Observatories," *Proc IRE* 50 (November 1962): 2216-24. Of course the U.S. had already conducted Project Argus, technically as part of the International Geophysical Year. The two IGY satellites, *Explorer* and *Vanguard*, were utilized for data collection and the data was subsequently made public. See Hanson W. Baldwin, "U.S. Atom Blasts 300 Miles Up Mar Radar, Snag Missile Plan," *New York Times*, March 19, 1959; "Excerpts from the Defense Department's News Conference on the Project Argus," *New York Times*, March 20, 1959; House Committee on Science and Astronautics, *Nuclear Explosions in Space*, 86th Congress, 1st sess., 1959, Committee Print.

33 Prelinger, *Another Science Fiction*, 198-203.

34 Warren Hanford and Barbara G. Bell, *Modernist Themes in New Mexico: Works by Early Modernist Painters* (Taos, NM: Gerald Peters Gallery, 1989); *Art and the Atom: An Exhibition of Contemporary Art Used in Scientific Advertisements* (Taos, NM: Stables Art Gallery, n.d. [c. 1963]), catalog of an exhibition curated by Robert Meier, assistant personnel director in charge of recruitment, Los Alamos Scientific Laboratory.

35 Prelinger, *Another Science Fiction*, 198-203.

## CHAPTER NINE: BIONICS, A PROLOGUE TO TRANSHUMANISM

1 Martin Caidin, *Cyborg* (New York: Warner, 1972).

2 Kim Tingley, "The Body Electric," *New Yorker*, November 25, 2013, 78-86.

3 Schaut, *Robots of Westinghouse:* 25-29.

4 Keith Massie and Stephen D. Perry, "Hugo Gernsback and RadioMagazines: An Influential Intersection in Broadcast History," *Journal of Radio Studies* 9 (Summer 2002): 264-80.

5 Henri Jeryan Puzant, "High-Speed Radio Production," *Electronics*, December 1935, 21-25.

6 Editorial, *Electronics*, June 1935, 193.

7 "Electronic Robot 'Greeter' at New Franklin Institute," *Electronics*, June 1934, 187.

8 Elektro appears in the sponsored film *The Middleton Family at the New York World's Fair* (Audio Productions, Inc. for Westinghouse, 1939), https://archive.org/details/middleton_family_worlds_fair_1939 (Internet Archive, Prelinger Archives collection) (accessed June 2012).

9 "New Portable Electro-cardiograph," *Journal of the AIEE* 49 (August 1930): 661.

10 Norbert Wiener, *Cybernetics: Or, Control and Communication in the Animal and Machine* (Cambridge, MA, and Paris: MIT Press, 1948).

11 Wiener, *Cybernetics*, 185-90.

12 Fred Turner situates Wiener's cybernetic theory within a total project of the formulation of the postwar "democratic personality." See Turner, "The Cold War and the Democratic Personality," in *The Democratic Surround: Multimedia and American Liberalism from World War II to the Psychedelic Sixties* (Chicago: University of Chicago Press, 2013), 151-80.

13 Grace M. Hopper, "Compiling Routines," *Computers and Automation* 2 (May 1953): 1-4.

14 "Remarks by M. J. Kelly, President, Bell Telephone Laboratories, Inc., before Bell System Lecturer's Conference," October 2, 1951, AT&T Archives and History Center.

14 W. L. Everitt, "Let Us Redefine Electronics," *Proc IRE* 40 (August 1952): 899.

15 *"Metal Brains" for Space Age Rockets* (Detroit: General Motors, 1961). The booklet describes analog gyroscopes used in attitude stabilization systems for missiles and rockets.

16 John Diebold and Associates, "Annual Computer Census," *Computers and Automation* 7 (May 1958): 8-9.

17 The earliest published use of the term "bionics" is in L. M. Butsch, Jr., "Bionics—A New Systems Technology," *Waveguide* (Dayton Section IRE publication), August-September 1960.

18  L. M. Butsch, Jr., and C. W. Gwinn, "Bionics—Status and Plans," *IEEE Transactions on Military Electronics* 7 (Summer 1963): 261-66.

19  William E. Bushor, "The Perceptron: An Experiment in Learning," *Electronics*, July 22, 1960, 56-59. Also Bushor, "Model Perceptron Makes Debut," *Electronics*, June 24, 1960, 43.

20  Mikel Olazaran, "A Sociological Study of the Official History of the Perceptrons Controversy," *Social Studies of Science* 26 (August 1996): 611-59.

21  J. C. R. Licklider, "Man-computer Symbiosis," *IRE Transactions on Human Factors* HFE-1 (March 1960): 4-11.

22  J. A. N. Lee, J. McCarthy, and J. C. R. Licklider, "The Beginnings at MIT," *IEEE Annals of the History of Computing* 14, no. 1 (1992): 18-54.

23  Manfred E. Clynes and Nathan S. Kline, "Cyborgs and Space," *Astronautics* 5, no. 9 (September 1960): 26-27, 74-76.

24  Alfred N. Goldsmith, "Electronics and Medicine," *Electronic Age* 25 (Winter 1966): 26-29.

25  I'm paraphrasing the science fiction writer William Gibson, who famously observed (many times, in many contexts) that "the future is already here, it's just not evenly distributed."

26  Emily McVarish, personal correspondence, May 31, 2014.

27  See Robert Poole, *Earthrise: How Man First Saw the Earth* (New Haven and London: Yale University Press, 2008).

28  See Edgar Zilsel, *The Social Origins of Modern Science* (Boston: Springer Verlag, 2003), a collection of essays written between 1911 and 1944, for a comprehensive analysis of the relationship between science and society.

29  See Jennifer Gabrys, *Digital Rubbish: A Natural History of Electronics* (Ann Arbor, MI: University of Michigan Press, 2011).

30  Iain Boal, personal communications, September 2013.

31  Kevin Kelly, *What Technology Wants* (New York: Penguin, 2010).

32  Gabrys, *Digital Rubbish*, 74-100.

33  Katherine McFadden, "Handsewn Computing: Women's Hobbies, Needlework, and Computer Electronics." Paper presented at Society for the History of Technology annual conference, Dearborn, MI, November 2014.

# REPOSITORIES CONSULTED

California College of Art Library
Denver Art Museum Library
Hagley Museum and Library
Lincoln Laboratory Archives Center
MIT Library
NASA History Archives
National Air and Space Museum Archives
National Museum of American History Archives
Prelinger Library and Archives
Schenectady Museum and Planetarium, General Electric Archives
University of California, Santa Cruz, Science and Engineering Library
University of Illinois Library and Archives

## DIGITAL REPOSITORIES

Computer History Museum Digital Archive
CyberneticZoo.com
IEEEXplore Digital Archive
Internet Archive
Ubuweb

# REFERENCES FOR FIGURES

All figures are magazine advertisements except where noted. Many of the advertisements that appear in this book were originally published several times across a range of magazines and journals, and even across spans of time that extended to years. The citations given below refer to the one best available copy of an advertisement that was scanned for this book. The citations do not always refer to the earliest published instance of an advertisement, though they do where possible. All figures are from the author's collection except where noted.

i.1 General Motors, *New World of Electronics*. Booklet with text by Martin Mann. Detroit: General Motors, 1964 (front cover).

i.2 Playing card. Schenectady, NY: General Electric, c. 1917. Adapted from artwork originally created for General Electric by Maxfield Parrish.

i.3 American Telephone and Telegraph Company. *Suburban Life* 13 (October 1911): 217.

i.4 Radio Corporation of America. *Scientific American* 203 (September 1960): 14.

i.5 Radio Corporation of America. *Scientific American* 203 (August 1960): 145.

i.6 "Electronics: Techniques for a New World 1: A Fortune Collage," designed by Vardø in collaboration with Warren Stokes and Antonio Petrucelli; with photographs by Fritz Goro, Frederic Lewis, Arthur Griffin, and Frank Scherschel. *Fortune* 14 (July 1943): insert.

i.7 Burroughs Corporation. *Aviation Week* 71 (November 2, 1959): 73.

i.8 Boeing Corporation. *Electronics* 34 (April 21, 1961): 37.

i.9 Allied Radio, *Catalog No. 124, 1951*. Chicago: Allied Radio Company, 1951. Cover illustration courtesy of Radio Corporation of America.

i.10 *Electronics* 32 (November 27, 1959): cover.

i.11 Institute of Radio Engineers. *Proceedings of the IRE* 49 (May 1961): 173a.

i.12 *Proceedings of the IEEE* 46 (May 1968): cover.

## CHAPTER ONE

1.1 General Electric, *Electronics—A New Science for a New World*. Booklet designed by Herbert Bayer. Schenectady, NY: General Electric, 1942: front cover.

1.2  Allied Radio, *Catalog No. 112, 1947*. Chicago: Allied Radio Company, 1947. Cover illustration courtesy of Electronics Department, General Electric Company (adapted from a design by Herbert Bayer).

1.3  Delco Radio Division of General Motors, with artwork by Harold Flucke. *Aero-Digest* 48 (January 1, 1945): 47.

1.4  Harold G. Bowen, *The Edison Effect*. Newark, NJ: The Thomas Alva Edison Foundation, 1951: front cover.

1.5  Lockheed. *Electronics* 13 (March 27, 1959): 26-27.

1.6  Merle Duston, *Electronics for Beginners*. Booklet cover design by Al Scott. North Canton, OH: Hoover Industries, 1946.

1.7  Concord Radio Corporation, *Catalog No. 95, 1952*. Chicago: Concord Radio Corporation, 1952.

1.8  General Electric, *Your Coming Radio as Forecast by General Electric*. Booklet. Schenectady, NY: General Electric, 1944: front cover.

1.9  International Business Machines. *Scientific American* 181 (December 1949): inside front cover.

1.10 Remington Rand, *Remington Rand Presents the Electronic Era for Business with UNIVAC*. Booklet. New York: Remington Rand, c. 1951: front cover. Source: Hagley Museum and Library.

1.11 Bankers Trust Company. *Business Week* 1098 (September 16, 1950): 1.

1.12 Continental Electronics Company. *Proceedings of the IRE* 49 (April 1961): 81a.

## CHAPTER TWO

2.1  The Ralph M. Parsons Company. *Scientific American* 205 (July 1961): inside back cover.

2.2  Tung-Sol Electric Incorporated. *Electronics* 33 (February 12, 1960): 123.

2.3  The Budd Company, with artwork by Raul Mina Mora. *Business Week* 1520 (October 18, 1958): 115.

2.4  Jet Propulsion Laboratory. *Proceedings of the IRE* 46 (October 1958): 143a.

2.5  Radio Corporation of America. *Electronics* 32 (November 20, 1959): back cover.

2.6  Allied Radio, *Catalog No. 111, 1946*. Chicago: Allied Radio Company, 1946.

2.7  Allied Radio, *Catalog No. 117, 1949*. Chicago: Allied Radio Company, 1949. Cover illustration courtesy Radio Corporation of America.

**2.8** Radio Corporation of America, with artwork by Arthur Lidov. *Proceedings of the IRE* 38 (February 1950): 104a.

**2.9** Allen B. Du Mont Laboratories, Inc. *Business Week* 1303 (August 21, 1954): 113.

**2.10** Varian Associates (detail). *Scientific American* 205 (August 1961): 63.

**2.11** Marquardt Corporation. *Aviation Week* 73 (December 12, 1960): 2.

**2.12** Stromberg-Carlson. *Business Week* 1478 (December 21, 1957): 107.

**2.13** Lafayette Radio, *Catalog No. 590, 1959*. New York: Lafayette Radio Company, 1959.

**2.14** Radio Corporation of America. *Proceedings of the IRE* 49 (September 1961): 63a.

**2.15** International Telephone and Telegraph Corporation. *Scientific American* 205 (September 1961): 97.

### CHAPTER THREE

**3.1** Raytheon. *Scientific American* 202 (January 1960): 14.

**3.2** *Proceedings of the IRE* 43 (December 1955): front cover.

**3.3** International Business Machines. *Proceedings of the IRE* 47 (February 1959): 119a.

**3.4** Sandia Corporation. *Scientific American* 203 (September 1960): 269.

**3.5** Stromberg-Carlson. *Proceedings of the IRE* 47 (January 1959): 30a.

**3.6** General Electric, *Electronics—A New Science for a New World*. Booklet designed by Herbert Bayer. Schenectady, NY: General Electric, 1942: 8.

**3.7** Walter Murch for *Scientific American*. *Scientific American* 181 (December 1949): front cover.

**3.8** Sandia Corporation, with artwork by Thomas Holland. *Scientific American* 205 (November 1961): 161.

**3.9** Ken Staley for General Electric. Painting in oil, 1965. Source: Schenectady Museum of Innovation and Science archives.

**3.10** The Lincoln Laboratory, with artwork by Jacqueline Casey. *Proceedings of the IRE* 53 (November 1965): 23a.

**3.11** Cornell-Dubilier Electric Corporation. *Proceedings of the IRE* 33 (January 1945): facing 74a.

**3.12** Corning Glass Company. *Business Week* 1563 (August 15, 1959): 61.

### CHAPTER FOUR

**4.1** Melpar, Incorporated (detail). *Missiles and Rockets* 5 (June 1959): 27.

**4.2** Eagle-Picher. *Business Week* 1954 (July 31, 1954): 41.

**4.3**   Texas Instruments. *Electronics* 37 (September 7, 1964): 139.

**4.4**   Texas Instruments. *Business Week* 1303 (August 21, 1954): 50.

**4.5**   Clevite Transistor. *Electronics* 33 (November 18, 1960): 17.

**4.6**   General Transistor Corporation. *Electronics* 32 (March 13, 1959): 149.

**4.7**   Litton Industries. *Electronics* 33 (September 9, 1960): 103.

**4.8**   General Electric, *Controlled Rectifier Manual*. Booklet. Syracuse, NY: General Electric, 1960: front and back covers.

**4.9**   Kintel Electronics, with artwork by M. Canning. *Missiles and Rockets* 9 (August 14, 1961): 33.

**4.10**  See 4.1.

**4.11**  W. W. Lenz and W. W. Cook, *Transistor Fundamentals and Applications*. New York: Radio Corporation of America, 1958: front cover.

**4.12**  General Electric, *Tubes and Transistors: A Comparative Study*. Schenectady, NY: General Electric, c. 1960: front cover.

**4.13**  Tung-Sol Electric Incorporated. *Electronics* 34 (September 29, 1961): 21.

**4.14**  Lockheed. *Aviation Week* 71 (September 14, 1959): 138-39.

**4.15**  *Photofact Reporter* (June 1964): front cover.

## CHAPTER FIVE

**5.1**   The Budd Company, with artwork by Raul Mina Mora. *Business Week* 1500 (May 31, 1958): 14.

**5.2**   St. Regis. *Electronics* 33 (June 24, 1960): 22.

**5.3**   Radio Corporation of America, *Printed Circuit Servicing Techniques*. Booklet. New York: Radio Corporation of America, 1960: front cover.

**5.4**   *Radio-Electronics* 29 (April 1958): front cover.

**5.5**   Ulano. *Electronics* 39 (June 13, 1966): 27-28.

**5.6**   *Space/Aeronautics* magazine, with artwork by J. M. Barton. *Industrial Marketing* 45 (June 1960): 106.

**5.7**   *Radio-Electronics* 30 (December 1959): front cover.

**5.8**   International Business Machines. *Electronics* 32 (April 24, 1959): 103.

**5.9**   The Martin Company, *The Story of a Giant*. Booklet designed by Willi Baum. Denver, CO: The Martin Company, 1961: 9. Source: Willi Baum collection.

**5.10**  Lockheed. *Electronics* 40 (April 3, 1967): 209.

**5.11**  General Electric. *Business Week* 1546 (April 18, 1959): 7.

**5.12**  General Electric. *Business Week* 1502 (June 14, 1958): 7.

**5.13**  International Resistance Company. *Electronics* 33 (September 16, 1960): 97.

**5.14** Pulsa. State of the Art Voltage-Controlled Sine-Wave Generator. Multi-media artwork, 1968. Source: David Rumsey/Pulsa.

**5.15** The Lincoln Laboratory, with artwork by Jacqueline Casey. *Proceedings of the IEEE* 63 (June 1975): back cover.

## CHAPTER SIX

**6.1** First National City Bank of New York, with artwork by Walter Murch. *Business Week* 1565 (August 29, 1959): 100.

**6.2** *Data Processing for Management* 6 (April 1964): front cover.

**6.3** Westinghouse (detail). *Electronics* 34 (August 11, 1961): 193.

**6.4** Laboratory for Electronics. *Proceedings of the IRE* 48 (June 1960): 47a.

**6.5** Litton Industries. *Proceedings of the IRE* 44 (November 11, 1956): 5.

**6.6** Lockheed. *Aviation Week* 71 (November 9, 1959): 60–61.

**6.7** Paul Rand, *Thoughts on Design*. New York: Wittenborn & Co., 1951: back cover (detail).

**6.8** *St. Louis Post Dispatch*. Advertising insert. St. Louis, MO: *St. Louis Post Dispatch*, 1927.

**6.9** Bank of America, *Payroll*. Booklet designed by Willi Baum. San Francisco: Bank of America, 1961: front and back covers. Source: Willi Baum collection.

**6.10** Remington Rand. *Business Week* 1316 (November 20, 1954): 151.

**6. 11** Tung-Sol Electric Incorporated. *Scientific American* 205 (October 1961): 125.

**6.12** Amphenol-Borg Electronics. *Electronics* 34 (March 17, 1961): 18.

**6.13** Controls Company of America. *Electronic Industries* 20 (August 1961): 139.

**6. 14** P. R. Mallory and Company. *Business Week* 1749 (March 9, 1963): 143.

**6.15** Underwood. *Business Week* 1751 (March 23, 1963): 23.

**6.16** General Time. *Business Week* 1628 (November 12, 1960): 137.

**6.17** Giannini Controls Corporation. *Electronics* 39 (June 13, 1966): 80.

**6.18** Bank of America, *Payroll*. Booklet designed by Willi Baum. San Francisco: Bank of America, 1961: unnumbered interior page. Source: Willi Baum collection.

## CHAPTER SEVEN

**7.1** Burroughs Corporation. *Missiles and Rockets* 5 (January 26, 1959): 35.

**7.2** Burroughs Corporation. *Business Week* 1573 (November 28, 1959): 87.

**7.3** Burroughs Corporation. *Missiles and Rockets* 5 (June 29, 1959): 23. Source: New York Public Library.

7.4 International Telephone and Telegraph Corporation, with artwork by Steve Chan. *Proceedings of the IRE* 48 (June 6, 1960): 131a.

7.5 Royal McBee. *Scientific American* 205 (September 1961): 25.

7.6 Royal McBee. *Business Week* 1659 (June 17, 1961): 86.

7.7 National Cash Register. *Scientific American* 206 (February 1962): 119.

7.8 Marquardt Corporation. *Aviation Week* 73 (September 19, 1960): inside front cover.

7.9 International Business Machines. *Proceedings of the IRE* 48 (November 1960): 103a.

7.10 *Proceedings of the IRE* 47 (October 1959): front cover.

7.11 International Business Machines. *Proceedings of the IRE* 49 (September 1961): 99a.

7.12 Bell Telephone Laboratories, *Computer Speech*. 45 RPM record, 1963.

7.13 Burroughs Corporation. *Data Processing* 3 (April 1961): 12-13.

7.14 The Lincoln Laboratory, with artwork by Jacqueline Casey. *Proceedings of the IEEE* 54 (September 1966): 19a.

7.15 The Lincoln Laboratory, with artwork by Jacqueline Casey. *Proceedings of the IEEE* 51 (October 1963): 88a.

7.16 International Business Machines. *Electronics* 32 (March 13, 1959): 300.

7.17 The Martin Company, with artwork by Aldcrofft. *Missiles and Rockets* 7 (September 5, 1960): 21.

7.18 Raytheon Company. *Electronics* 38 (October 18, 1965): 142.

## CHAPTER EIGHT

8.1 International Business Machines. *Scientific American* 204 (March 1961): 90.

8.2 *Electronics* 34 (November 17, 1961): front cover.

8.3 *Electronics Today* 1 (November 1959): front cover.

8.4 Allied Radio, *Catalog No. 160, 1957*. Chicago: Allied Radio Company, 1957.

8.5 Bell Telephone Laboratories. *Business Week* 1520 (October 18, 1958): 3 (detail).

8.6 Bell Aircraft. *Missiles and Rockets* 6 (May 2, 1960): 2 (detail).

8.7 Marquardt Corporation, with artwork by Ken Smith. *Proceedings of the IEEE* 51 (December 1963): 79a.

8.8 The Martin Company, with artwork by Red Gates. *Scientific American* 204 (February 1961): 129.

8.9 Philco. *Aviation Week* 15 (October 14, 1957): 137.

8.10 Honeywell. *Missiles and Rockets* 8 (March 20, 1961): 22.

8.11 General Electric. *Aviation Week* 71 (October 19, 1959): 120.

**8.12** Astrodata. *Electronics* 15 (July 26, 1965): 11.

**8.13** Radiation Incorporated, with artwork by Paul Calle. *Scientific American* 206 (April 1962): 83.

**8.14** Radiation Incorporated, with artwork by Paul Calle. *Scientific American* 206 (December 1962): 149.

**8.15** Radiation Incorporated, with artwork by Paul Calle. *Astronautics and Aeronautics* [then *Astronautics and Aerospace Engineering*] 1 (August 1963): 14.

**8.16** Radio Corporation of America. *Missiles and Rockets* 9 (November 13, 1961): 26-27.

**8.17** Marquardt Corporation, with artwork by Ken Smith. *Aviation Week* 74 (June 26, 1961): 2.

**8.18** Los Alamos Scientific Laboratory, with artwork by Emil Bisttram. *Nucleonics* 18 (December 12, 1960): 47.

**8.19** The Martin Company, with artwork by Willi Baum. *Missiles and Rockets* 10 (March 6, 1961): 30.

## CHAPTER NINE

**9.1** Brown Boveri Company. *Electrical World* 15 (April 9, 1962): 69.

**9.2** Bendix. *Aviation Week* 72 (May 11, 1960): 9.

**9.3** American Federation of Musicians, with artwork by Jesús Helguera. *The Literary Digest* (August 16, 1930): 3.

**9.4** Radio Corporation of America, unknown 78 RPM record (interior sleeve), 1934. Source: Eric Fischer collection.

**9.5** J. Richard Johnson, *How to Troubleshoot a TV Receiver*. 2nd ed. New York: John F. Rider Publisher, Inc., 1958: front cover.

**9.6** Keuffel and Esser Company. *Missiles and Rockets* 9 (December 18, 1961): 8.

**9.7** Gilfillan. *Aviation Week* 70 (March 16, 1959): 20.

**9.8** Edmund C. Berkeley, *Giant Brains, or, Machines That Think*. New York: John Wiley & Sons, 1949: dust jacket detail.

**9.9** United States Rubber, with artwork by Gaston Sudaka. *Business Week* 1457 (August 3, 1957): 127.

**9.10** Laboratory for Electronics. *Aviation Week* 72 (April 25, 1960): 137.

**9.11** Fairchild. *Aero-Digest* 73 (July 1956): 39.

**9.12** The Martin Company. *Proceedings of the IRE* 46 (March 1958): 244a-245a.

**9.13** Vitro Electronics. *Electronics* 35 (August 3, 1962): 13.

**9.14** *Electronics* 35 (February 9, 1962): front cover.

**9.15** Cornell Aeronautical Laboratory. *Proceedings of the IRE* 46 (November 1958): 133a.

**9.16** *Proceedings of the IRE* 47 (November, 1959): front cover.

**9.17** The Martin Company. *Electronics* 31 (June 27, 1958): 22-23.

**9.18** Radio Corporation of America. *Scientific American* 210 (March 1964): 75.

# ACKNOWLEDGMENTS

My first thanks are to my agent, Mary Evans, and to my editor at Norton, Brendan Curry. This book would not have reached tangible form without either of them and their respective tireless and boundlessly creative efforts on its behalf. I am grateful to both of them. Mitchell Kohles and Sophie Duvernoy at Norton were each very helpful in seeing the book through production, as were Anna Oler, Nancy Palmquist, Leslie Huang, and Charles Newman, and the text was clarified by the keen pen of Allegra Huston. The draft manuscript was considerably improved by critical feedback from Emily McVarish and from Paul Ceruzzi, each of whom was generous enough to examine the work through the eyes of their respective disciplines—the history of graphic design and the history of computing. Any errors that remain are my own.

Research for this book was supported in part by a grant from the Seed Fund, and in part by the Hagley Library and Museum travel research fund. Professional research assistance was provided by Christina Linden, with additional research assistance by Emma Hurst and Heather Jovanelli. The following individuals provided special assistance to this project as informants, subject-matter experts, materials scouts, and contributing artists: Craig Baldwin, Willi K. Baum, Patricia Belen and Greg D'Onofrio, Timothy Caldwell, Yves Feder, Eric Fischer, Dr. Hart Lidov, Nora Ligorano and Marshall Reese, Beatrice Murch, Walter Murch, Elizabeth Staley, and David Rumsey. Willi K. Baum in particular opened his life to many hours of interviews over the years, giving me a privileged look at the life of a Swiss modernist émigré graphic artist. Scholars in affiliated fields who offered assistance include Kevin Hamilton and Ned O'Gorman at the University of Illinois, where I visited in 2011, as well as David Cox and Molly Hankwitz, Iain Boal, Katherine McFadden, and Lisa Gitelman. Others who pointed to helpful resources along the way include Ron Kihara, Jon Leidecker, Roger Manley, Caroline Martel, Steve Massey, Milton Minneman, and Martin Thibodeau.

The Prelinger Library community of volunteers, regulars, and visitors formed the nucleus of the working environment for this book. The shared research environment of our weekly public hours is the rhythmic pulse behind all the work that we do. Freya Channing was our anchor in the library while this book was being written, and her librarianship often made it possible for work to be done on the manuscript even during open hours. Charlie

Macquarie's librarianship helped support work on the manuscript in its later stages. The artists group Futurefarmers invited me to collaborate with them in 2012, a gesture that contributed materially to this book by causing me to think much more deeply about the concept of scale than I would have done otherwise. The Other Cinema screening series offered opportunities to present work-in-progress talks; my thanks to Craig and to the entire ATA community. Collaborative opportunities offered by BoingBoing and Nerd Nite also helped to shape the focus of this book, and O'Reilly Media offered me the first opportunity to present some of this material back in 2009. Membership in the extended Exploratorium community provided an added supportive context for this work through its duration. My thanks to Susan Schwartzenberg at the Museum for her help and understanding along the way.

At the Hagley Museum and Library, where I was fortunate to spend an entire week in 2011, I am grateful for the support of Roger Horowitz and Carol Lockman at that institution's Center for Business, Technology and Society. My research at the Hagley was facilitated by the librarians and archivists Lucas Clawson, Max Moeller, Lynn Catanese, and Chris Baer. Other librarians and archivists who facilitated research at institutional repositories include Elizabeth Suckow, Colin Fries, and John Hargenrader at NASA headquarters; Elizabeth Borja and Allan Janus at the Smithsonian National Air and Space Museum; Jim Roan and Michael Hardy at the Smithsonian Museum of American History; Tamar Granovski at Lincoln Laboratory; and Chris Hunter at the Schenectady Museum of Innovation and Science (General Electric archives). Rick Prelinger conducted research on behalf of this project at the Denver Art Museum (Herbert Bayer archives), where he was helped by the archivists Renée B. Miller and Gwen Chanzit. Deborah Douglas at the MIT Museum and George Kupczak at the AT&T Archives and History Center responded helpfully to email queries.

In addition to the research institutions listed above, I made considerable use of digital archives. Among those, the Internet Archive deserves special appreciation. It is a digital utopia where everything is free and fully downloadable, and is also a friend and ally here in San Francisco. Ubuweb is another digital utopia that was helpful for this project.

This book was written with the moral support of many family members and friends. My dad, Robert Shaw, a former aviator who is also an artist, read chapter drafts and offered helpful critical feedback. My loving mom is Kimberly Stevens. The Prelingers let me join them: Ernst and Rosemarie, and sisters Polly Prelinger and Jane Prelinger. Special thanks to the art historian Elizabeth Prelinger who was also a research advisor, and to Stephen, Benjy, and Frederick Messner for logistical support during research trips to Washington DC. I appreciate the moral support of Barbara Shaw and her family, of the Root, Maddox, and Welters families; of Linda Kashin and Maggie Hadley and their families, and of Glenna Allee, Nancy Appel, Amy Balkin and Josh On, Kimiko Baum, Summer Brenner, Gwyneth Cravens, Alicia Curtis and Kathy Pratt, Sandi Deckinger, Katrina Lamberto Elsheimer and Skip Elsheimer, Severine von Tscharner Fleming, Kerry Laitala, Caroline Martel, Susan McCarthy, the Parker-Yokoyama family, Eric Peterson, Deborah Rodgers, Abby Smith Rumsey, Scott Stark and Kathleen Tyner, Dan Wilson, and the House of Zoka, among many others I am unable to name individually.

For general advice and encouragement at an early stage of the project I wish to thank Frank Winter and to remember Frederick Ordway. My thanks also to the staff at the NASA History office and the curators at the Smithsonian National Air and Space Museum, all of whom have been welcoming and helpful to a citizen historian. Spouse Rick Prelinger contributed research and logistical assistance, and moral support, on a scale that dwarfs the contributions of all others. Research on this book was originally prompted in part by my fascination with the historical radio ephemera that he has been collecting since childhood. This then, like everything that I do, is for you, Z.

# INDEX

Page numbers in *italics* refer to illustrations.
Page numbers beginning with 222 refer to endnotes.

abacus:
    as digital computer, 126-28, 231-32
    as graphic symbol, 124, 125, 127-
        28, *130, 140,* 146-47, *218*
abstraction, in art, 95, 96, 110-11,
    113-14, 203
adding machines, 151
Advanced Research Projects Agency
    (ARPA), 162, 186
advertising:
    business-to-business, 16, 22, 173
    photography in, 141
    *see also* graphic art; trade
        literature; *specific publications*
*Aero-Digest, 207*
aerospace industry, 178
Agena rocket, 177, *177*
"age of information," 143, 161
AI. *See* artificial intelligence
Air Force, U.S., 177
Albany Institute of History and Art, 81
ALGOL, 154, 159
algorithms, 153
Allied Radio catalog, *23, 32,* 51, *51, 52,*
    55, *175*
alternating current, 13
American Federation of Musicians,
    199, *199*
American Indian Movement, 117
American Standards Association, 95,
    138-39
Amphenol-Borg, *136*
amplification, of signals, 86, 88
AM radio, 29-30, 55-56

*Angers of Spring, The* (Whitehill), 103
Antheil, George, 136
*Anthology of Concrete Poetry*
    (Williams, ed.), 149
Anthropocene, 193
Apollo, Project, 168-69, 171, 178, *183,*
    185, *186,* 195
Apollo 8, 220
Apollo 11, 185
Armstrong, Edwin, 29-30
Army, U.S., 178, *182*
Army Signal Corps, U.S., 49, 73, 105,
    108, 213
ARPA (Advanced Research Projects
    Agency), 162, 186
art:
    abstract, 95, 96, 110-11, 113-14,
        203
    conceptual, 119, 141-42
    Constructivism, 15, *70, 147,* 148,
        *182*
    CRTs as media for, 56-57
    Cubism, 95, 148
    dialogue between science and, 20,
        21, 22-23
    dialogue between technology and,
        20-25, 27, 96, 148, 163, 173, 219
    Expressionism, 190
    Futurism, 15, 95, 148
    intangible made visible by, 11, 12
    modernism, 16, *21,* 23-25, *83,* 89,
        90, 113, 119, 136-37, 201-3, 215,
        216, 219
    neoclassicism, 75

art (*continued*)
  Romanticism, 95
  surrealism, 15, *52*, 63, 75, *202*, 203
  video and time-based media in, 56
  *see also* graphic art
*Art and Science*, 76
artificial intelligence (AI), 41, 126, 155,
    157, 158, 199, 203, 204, 205, 209,
    213, *213*, 215, 236
"Art in Science" (exhibition), 81
art nouveau, 95
Arts and Crafts movement, 95
ASCII (American Standard Code for
    Information Interchange), 138-39
Aspen Design Institute, 32
Association for Computing
    Machinery, 208
Astrodata, *183*
*Astronautics and Aeronautics*, *186*
astronauts, heroic image of, 182, *183*,
    191, 195, 216
astronomy, 55
  radio, 50
AT&T, 13, 25
  *see also* Bell Laboratories
Atlas rocket, 113, 180, 186
atom, Rutherford-Bohr model of,
    35-36, 37, *43*, 70, 74, 171, *213*
  as icon of electronics industry, 31,
    37, 42-43, 51, *52*
  as symbol of nuclear threat, 41-42
atomic bomb, 42, 151
atomic clock, 74
*Atomic Ranch*, 42
atoms:
  lattices of, 83
  nuclear force in, 69
audion tube, 35, *100*
automation, 15, 123-26, 141, 232
  pushback against, 143
Auto-Sembly, 108-9
aviation communications, 33-34
*Aviation Week*, *21*, *58*, *100*, *130*, *153*,
    155, *180*, *182*, *188*, *198*, *203*
avionics, 34, 177, 217

Babbage, Charles, 134, 137
Ballistic Missile Early Warning
    System, 64-65
ballistic trajectory calculations, 126
Banker's Trust, *42*
Bank of America, *134*
Bardeen, John, 86-87, 88, 89, 117
Barton, J. M., 108, *109*
*Bash* (Paolozzi), 81
Bauhaus School, 15, 19, 29, 95, 148,
    200
Baum, L. Frank, 198
Baum, Willi, *111*, 113, 114, *134*, *193*
Bayer, Herbert, 29, 30, 32, 37, 39, *73*,
    74, 148, 223-24
  as both fine and commercial artist,
    33
  "Earth Bulb" of, *30*, 31, *32*, 33, 36,
    51, 163, 202
beacon-assisted flight, 37
Bell, Alexander Graham, 13
Bell Aircraft, 176-77, *177*
Bell Laboratories, 13, *14*, 18, 20, 21,
    33, 40, 73, 76, 77, 79, 99, 108,
    126, 131, 158-59, *158*, *175*, 203-4,
    213
  satellite programs of, 183-85
  transistor developed by, 86-87,
    *86*, 88, 89, 93, 117, 131, 208
Bell System, 88
*Bell Telephone Magazine*, 123
Bendix, *198*
Bentham, Jeremy, 65
Berkeley, Edmund C., 137, *203*, 208
"Bicycle Built for Two," 258-59
binary data systems, 39, 126, 133, 137,
    140, 157, 161
biomorphic modeling, of computer
    memory, 215
bionics (biomedical electronics), 24,
    195-220, *211*, *213*, *215*
  electronic circuits in, 197
  science fiction and, 197, 198, 217
  use of term, 211, 213, 217
Bisttram, Emil, *189*, 190-91

bits, coining of term, 133
Bochat, Catherine, *107*
Boeing, *22*, 176
Bohr, Niels, 35–36, 171
Braille, 139
brain:
    electrical signals in, 200–201
    electronic, *see* electronic brain
"Brain as a Computing Machine, The"
    (McCulloch), 207–8
Brattain, Walter, 86–87, 88, 89, 117
Brown Boveri Company, *196*
Budd Company, *47*, *104*, 115
Burroughs Adding Machine Company,
    151
Burroughs Corporation, *21*, *146*, *147*,
    *148*, 151, *159*, 173, 178, 180, 186
    B 5000 computer of, 159, 161, 178,
      180
    Datatron 205 computer of, *146*,
      *147*, 151
*Business Week*, 26, 42, *42*, *47*, *55*, *60*,
    61, *83*, *86*, *124*, *139*, *140*, *147*, *150*,
    *175*, *180*
bytes, coining of term, 133

Caidin, Martin, 195, 211
Caldwell, Orestes H., 26–27
Caldwell-Clements, 27
California Museum of Science and
    Industry, 224
Calle, Paul, *184*, 185, *185*, *186*
Canning, M., *94*
capacitors, 83, 88, 103, *106*
Čapek, Karel, 198
Carnegie Corporation, 213
car radios, 88
Casey, Jacqueline, 79, *80*, 81, *120*, 121,
    *160*, 162–63, *165*, 166, 169, 237
cathode-ray tubes (CRTs), 41, 45–65,
    91, *149*, *158*, 200
    as artistic media, 56–57
    as computer input/output devices,
      60–61, 63–65, 157, 166, *168*
    habituation to viewing of, 46, *47*

    as information storage devices,
      59–60
    language and, 57–59, 147
    pressure-sensitive screens of, 61
    schematic symbol for, 98
    television and, 50–52, *52*, 55–56, *55*
    wall-sized projections of, 63–65, 77
    wartime applications of, 49
census, U.S., of 1890, 134–35, 232
Center for Advanced Visual Studies,
    20
Ceruzzi, Paul, 166
Chan, Steve, *64*, 65, *149*
chaos theory, *150*
character recognition, 57, 147, 155,
    157–58, *157*, 213
Charactron, *60*, 61
circuit boards, 24, 103–21, *110*, *113*, 123
    brain equated to, 105, 106
    craft-based techniques and, 106–
      8, 120–21
    as design motif, 105, 107, 110–11,
      119, *148*
    hand-assembled, 104–5
    master drawings for, 107
    miniaturization of, 177
    printed, *see* printed circuits
    television and, 110
    *see also* integrated circuits
circuits, electronic, 88
    bionic, 197
    schematic symbols for, 92–96, *92*,
      *93*, 98, *100*, 101, *148*, 149
Clevite Transistor, 90, *91*
clocks:
    atomic, 74
    quartz, 73–74, 86
Clynes, Manfred, 216–17
Coates, Robert, 81
COBOL, 154–55
Coiner, C. T., 31
Cold War, 41, 63, 64, 65, 77, 93–95,
    96, 114–15, 151, 162, 166–67, 171,
    175, *182*, 188–89, 191, *196*, 205,
    213, 218, 219

collage, 19, 215-16
Colossus computer, 39
commercial artists:
    anonymity of, 33
    see also graphic art
communication theory, 157, 203-4
Compass Island, USS, 198
compilers, 154
computation, mathematical, see
    mathematical computation
Computer Boys Take Over, The
    (Ensmenger), 152
Computer Professionals for Social
    Responsibility, 143
computer programmers, 133, 152-55,
    154, 163, 166
    women as, 25, 155, 235-36
computers, 29
    analog, 118, 126, 127, 231
    biomorphic memory in, 215
    brain equated to, 180, 206-8,
    209-10
    digital, 118, 123-43
    as dominant electronic
    technology, 166
    language and, see language,
    computers and
    mainframe, 11, 24, 39-41, 141-43,
    147, 161, 180
    merger of humans and, 215
    military origins of, 39
    monitors for, 45, 58, 63-65
    networks of, 10
    neural, 213
    printed circuits in, 109-10, 141
    space programs and, 177-78, 180,
    180
    tablet, 130
    transistors and, 99, 110, 141,
    145-46
    vacuum tubes and, 39-41, 110, 126
    weaving and, 121
computers, input/output systems for:
    CRTs as, 60-61, 63-65, 157, 166,
    168

magnetic tape as, 125, 140-41, 158
    see also paper tape; punched
    cards
computers, storage technology for,
    153
    CRTs as, 59-60
Computers and Automation, 155
Computers and People, 208
Computer Speech (record), 158
conceptual art, 119, 141-42
Concord Radio catalog, 36, 37, 37
concrete poetry, 149
Constructivist movement, 15, 70, 147,
    148, 182
Container Corporation of America,
    32
Continental Diamond Fibre, 104
Continental Electronics, 43
Control Data Corporation, 151
Controls Company of America, 138
Cooper, Muriel, 162
Cornell Aeronautical Laboratory, 213,
    213, 215
Corning Glass Company, 83
countdowns, 167, 167, 169
crafts, electronics and, 106-8, 120-21
Crane, Walter, 95
Crisis We Face, The (Steele and
    Kircher), 143
CRTs, see cathode-ray tubes
cryptography, 167-68
crystalline, use of term, 70, 71
crystallography, 77, 87
crystals and crystalline structures,
    67-83
    electromagnetic emissions from
    (piezoelectricity), 71, 72, 75-76,
    79
    graphic design and, 71, 76, 81, 83,
    90
    growing of, 72-73
    handedness of, 72
    as semiconductors, 82, 86-87
    structure of, 68-71, 68, 70, 76-77,
    80, 81

symmetry of, 70-71
synthetic, 75-76, 83
Crystals Design Project, 77
*Crystals Go to War* (film), 73
Cubism, 95, 148
Curie, Pierre and Jacques, 71
Cuthill, Elizabeth, 236
cybernetics, 203-9
    in warfare, 218-19
*Cybernetics* (Wiener), 204, 207, 208,
    209
*Cyborg* (Caidin), 195
cyborgs, 195, 211
    robots as mirror image of, 197-98
"Cyborgs and Space" (Clynes and
    Kline), 216-17

DARPA, *see* Advanced Research
    Projects Agency
*Datamation*, 141
data processing, *146*, 148, 149, 150,
    151, 152, 153, 159, 161
    automation of, 123-24
    electromechanical, 133
*Data Processing*, 141, *159*, 161
*Data Processing for Management*,
    125, *125*
Datatron 205 computer, *146*, *147*, 151
David, Edward E., 152
Dawn spacecraft, 188
De Blois, R. W., 79, 81
decision trees, 126, 157
Defense Department, U.S., Advanced
    Research Projects Agency of,
    162, 186
de Forest, Lee, 35, *100*, 228
dehumanization, technology as force
    for, 136
Delco Radio, *34*
dematerialization, 119
DEW (Distant Early Warning) Line,
    162, 166
difference engine, 134
*Difference Engine, The* (Sterling and
    Gibson), 130-31

Digitair, *107*
*Digital Apollo* (Mindell), 178
digitalization, 124, 125, 129, 135, 141,
    213, 231
diodes, *24*, 83, *87*, 88, *91*, *106*
distance, telephone and human
    experience of, 13, *14*
Douglas, 176
drone warfare, 191
*Dr. Strangelove*, 65
Druyan, Ann, 191
Dumont Laboratories, 55-56, *55*
Duston, Merle, 36
Dyna-Soar spaceplane, 176

Eagle-Picher, *86*
Eames, Charles and Ray, 224
"Earth Bulb," *30*, 31, *32*, 33, 36, 51,
    163, 202
*Earthrise*, 220
*Echo* satellites, 183-84, 185
*EchoStar XVI* satellite, 193
Eckert, J. Presper, 39-40, *40*, 41, 126,
    131, 133, *135*, 154, 233
ecological systems, 221, 232
Edison, Thomas, 13, 34-35
Edison effect, 34-35
Edison Mazda lightbulb, *12*, 13
Egbert (robot), 201
Einstein, Albert, 85-86, 171
*Electrical World*, *196*
electricity, 34-35
    in human body and brain, 200-201
    life and, 67-68
ElectroData Corp., 151
electromagnetism, 68-70
    atmospheric, 176
    in crystals, 71
electronic brain:
    circuit board as, 105, 106
    computer equated to, *180*, 206-8,
    209-10
    as graphic motif, 105, 106, *180*,
    200, *200*, 207, 209-10, *218*
    television compared to, *201*

*Electronic Engineering*, 207-8
*Electronic Industries*, 27, *138*
electronics:
   beginnings of, 12-14
   biomedical, *46*, 52
   coining of term, 217
   computers as dominant
      technology in, 166
   craft-based techniques and, 106-
      8, 120-21, 221
   as extensions of human
      sensorium, 24, 55, 74, 91, 191,
      197, 209, 216
   human figure as visual trope for,
      215-16
   human sense of scale expanded
      by, 146
   integration into everyday life of,
      86
   measurement and, 47-48, *47*
   micro-, 119-21
   miniaturization of, *see*
      miniaturization
   modernity and, 23, 86
   music and, 56-57, 59
   as "organic" vs. "mechanical," 23,
      72-75
   planetary model and, 220-21
   postwar boom in, 41, 87, 151
   redefining of, 208-9
   solar system as metaphor for, *11,
      12*, 13, 23, 31, 36, 37, 220
   space and, *see* space electronics
   surveillance and, 51, 63, 64, 65,
      114, 151, 162, 166
   vacuum tube as visual icon for, 36
   visibility of, 10
   wartime research in, 33-34, 36, 41,
      49, 73, 87
   waste processing of, 220-21, 232
   *see also specific technologies*
*Electronics*, 10, 15, *22, 24*, 26-27, *36,
   49, 56, 57, 59, 87, 91, 92, 100,
   101, 105*, 107-8, *108, 111, 112, 116,
   126, 136, 140*, 141, 149, 150, *168,*

   174-75, *174*, 183, 200-201, *211,
   213, 216*
"Electronics: A New Science for a
   New World" (GE booklet), 30-32,
   *30*, 36-37, *73, 74*
"Electronics: Techniques of a New
   World 1" (poster), 17-20, *18*, 216
*Electronics Engineering*, 177
*Electronics for Beginners*, *36*, 55
electronics industry:
   art-technology dialogue and,
      22-23
   business-to-business advertising
      by, 16, 22
   women in, 25, *82*
*Electronics Today*, 174, *174*
*Electronic Technician*, 27
electron microscope, 77
Electron Microscope Society of
   America, 81
electrons, 35, 69-70
   flow of, 10, 11, 24, 79, 82-83, 103
   invisibility of, 16
electrostatic storage tube, 59-60
Elektro (robot), 201
elements, 69
Emerson Electronics, 110
ENIAC, 39-41, 126, 141, 145, 154
Ensmenger, Nathan, 152
environmental movement,
   technology and, 220-21
ethylene diamine tartrate (EDT), 75-76
Europe, interwar exodus of artists
   and scientists from, 15-16, 24, 29,
   32, 190
Everitt, William L., 208-9
Experiments in Art and Technology
   (EAT), 20
Experiments in Disintegrating
   Language, 168
*Explorer* (IGY satellite), 175, 240
*Explorer II* (stratospheric balloon),
   175-76
Expressionism, 190
eye, persistence of vision and, 55

Fairchild Engine and Aircraft
    Corporation, *207*
Fairchild Semiconductor, 89, 117, 177
    Shiprock factory of, 117, 190
Faraday, Michael, 68, 69, 95
Farnsworth, Philo T., 50-51, 52
fax machines, 58
ferrites, 79
ferro-magnetic domains, 79, *80*, 81
Festival of Britain, 77
Fink, Donald G., 101
First National City Bank, *124*, 125
Flucke, Harold, *34*
Fly, James Lawrence, 45
FM radio, 29-30, 37, 41, 55-56, 223
FORTRAN, 154
*Fortune*, 16-20, *18*, 25-26, 36, 85, *180*
*Frankenstein, or the Modern
    Prometheus* (Shelley), 67, 198
Franklin, Benjamin, 34, 201
Franklin Institute, 201
*Future Shock* (Toffler), 143
*Futurist Manifesto*, 15
Futurist movement, 15, 95, 148

galaxies, 74
gamma-ray spectrometer, 190
Gates, Red, *178*
Gemini, Project, 168-69, 171, 172-73,
    180, 210
General Dynamics, 61, 72
General Electric (GE), *12*, 13, 18, 19, 21,
    25, *32*, 33, *38*, 79, *80*, *93*, 99, *113*,
    126, 150, 176, *182*
    "Electronics" booklet of, 30-32,
        *30*, 36-37, *73*, 74
    FM radio and, 29-30, 223
    in-house art department of, 79, 81
General Motors, *11*
General Time, *140*
General Transistor Corporation, *92*
Genuys, François, 145
germanium, 86-87, *86*, 89, 116
Germany, Nazi, 32
Gernsback, Hugo, *110*

Giannini Controls Corporation, *140*
*Giant Brains, or, Machines that Think*
    (Berkeley), 137, *203*, 208
Gibson, William, 130-31
Gilfillan, *203*
GL-880 transmitting tube, 30-31
Goldsmith, Kenneth, 169
Goro, Fritz, 19
graphic art:
    Bauhaus School and, 15, 19, 29, 95,
        148, 200
    circuit boards as motif in, 105, 107,
        110-11, 119, *148*
    crystalline structures and, 71, 76,
        81, 83
    depiction of information in, 142-43
    dialogue between technology and,
        20-25, 27, 96, 148, 163, 173, 219
    electronic brain as motif in, 105,
        106, *180*, 200, *200*, *207*, 209-10,
        *218*
    geometric form in, 95-96
    intangible made visible by, 11, 12
    integration of text and image in,
        148, 162-63
    mathematical symbols in, 146-47,
        148
    modernism and, *see* art,
        modernism and
    paper tape as motif in, 139, *139*,
        146-47, *149*, *158*, 159, *206*
    photography and, 141, 161, 219
    punched cards as motif in, 125,
        *125*, *134*, 148, 159
    robots and, 201-3, 205-6
    schematic symbols in, 92-96, *92*,
        *93*, 98, *100*, 101, *148*, 149
    science fiction and, 24, 174, 203
    Swiss tradition of, 162-63
    transistor and, 86, 89-96, *91*, *92*,
        *93*, 98
    typography in, *see* typographic
        art
    Viennese Secession and, 95
    *see also* art

*Graphic Science: The Magazine for Draftsmen*, 107
GrassMound (Bayer), 32
gravitational force, 69
Great Depression, 14, 52

Habbart, D. M., 158
HAL (char.), 258
Hamilton, Kevin, 56
handcraft, electronics and, 106–7, 221
handedness, 72
Harvard University, 41
Helguera, Jesús, 199, *199*
Herbert Televox (robot), 199
Hertz, Heinrich, 49
Hewlett-Packard, 89
Holland, Thomas, *79*
Hollerith, Herman, 134–35
Hollerith machine, 134–35
Honeywell, 150, 173, *181*
Hoover Industries, *36*
Hopper, Grace, 154, 208
Hughes Aircraft, *107*, 176
human body:
    electrical signals in, 200–201
    merger of computers and, 215–16
    as visual trope, 215–16
human sensorium, electronics as extensions of, 25, 55, 74, 91, 191, 197, 209, 216
human spaceflight programs, 168–69, 171, 172–73, 176, 178, 181–82, 185, 191, 195, 197

IBM, 18, 25, 38–39, *39*, 40, *69*, *111*, 128, 133, 150, *154*, 155, 157, *157*, *159*, 166–67, *166*, *172*, 173, 178, 186, 224
    608 computer of, 141
    700 series of, 145–46
    704 computer of, 145, 158, 159
    7090 computer, 145
ICBMs (intercontinental ballistic missiles), 151, 210
iconoscope, 51

ICs, *see* integrated circuits
*IEEE Transactions on Audio and Electroacoustics*, 118
*Industrial Marketing*, 22, 43, *69*, 108, *149*
inertial guidance systems, 177
information, graphic depiction of, 142–43
information overload, 143
information systems, 11, 232
information theory, 204, 205, 213
Institute for Advanced Study, 41
Institute of Radio Engineers (IRE), 95
    *see also Proceedings of the Institute of Radio Engineers*
integrated circuits (ICs), 101, 110, 116–19
    concerns about, 117–18
intercontinental ballistic missiles (ICBMs), 151, 210
International Electric Corporation, 65
International Geophysical Year (IGY), satellites and, 173, 175, 176, 240
International Resistance Company, *116*
International Telephone and Telegraph Corporation (ITT), *64*, 65, *149*
ion drives, 188
*IRE Transactions on Human Factors*, 27

Jacquard, Joseph Marie, 133, 135
Jet Propulsion Laboratory, *48*
*Journal of Applied Physics*, 19, 81

Kauner, H., 231–32
Kelly, Kevin, 220
Kelly, Marvin, 208, 209
Kepes, György, 20, 77, 79, 81
Keuffel and Esser Company, *202*
keyboards, *140*
Kilby, Jack, 116, 117
kinescopes, 51, *52*, 55, 65
kinetic sculpture, 203

Kintel, *94*
Kircher, Paul, 143
Kline, Nathan, 216-17
klystron tube, 19, *57*
Kubrick, Stanley, 65, 158, 159

Laboratory for Electronics, *128, 206*
Lafayette Radio catalogue, *61*, 63
Lang, Fritz, 198-99
Langevin, Paul, 71
language, computers and, 161
    character recognition, 57, 147, 155,
      157-58, 213
    concrete poetry and, 149
    CRTs and, 57-59, 147
    programming languages, 147, *150*,
      152-55
    synthesized speech, 7, 147, 152,
      155, 158-59, 213
*Language of Vision* (Kepes), 79
*Last Pictures, The* (Paglen), 193
lattices, of atoms, 83
Laue, Max von, 68, 69
launch countdowns, 167, *167*, 169
LED computer monitors, 45
LED technology, 45
*Leonardo*, 20
letterforms, *see* typography
Lewis, Frederic, 19
Licklider, J. C. R., 215
Lidov, Arthur, *52*
light, as both wave and particle, 68
lightbulb, 12, *12*, 13, 34, 56
    and human experience of time, 13
Ligorano/Reese, 121
Limited Nuclear Test Ban Treaty
    (1963), 190
Lincoln Laboratory, 67, 79, *80, 81,*
    *120*, 121, 157, *160*, 162-63, *165,*
    166, 178, 237
*Line and Form* (Crane), 95
Linotron typesetting machines, 61
Linotype-Paul, 61
Lippard, Lucy, 119
*Literary Digest*, 199, *199*

Litton Industries, *92, 129*
Lockheed, 35, *36, 100, 112,* 114, *130,*
    176
Lockheed Martin, 77, 79, 113, 114
logic gates, 126
Lorenz, Edward, *150*
Los Alamos Scientific Laboratory,
    *189*, 190
Lovelace, Ada, 137
Luddites, 232
Lukjanow, Ariadne, 235-36

machine languages, 152-53, 154
"Magic Brain," 200, *200*
magnetic domains, 79, *80*, 81
*Magnetic Domains* (Casey), 79, *80*, 81
magnetic tape, *125*, 140-41, *158*
magnetism, 68-69
magnetite, 79
Mahoney, Michael Sean, 231
mainframe computers, 11, 24, 39-41,
    141-43, 147, 161, *180*
Malina, Frank, 20
"Man-Computer Symbiosis"
    (Licklider), 215
Manhattan Project, 77
Mark series (computers), 41, 141
Marqatron, 189
Marquardt, *58, 153*, 155, *178, 188*, 189
Mars missions, 197, 218
Martenot, Maurice, 57
Martin Company, *111*, 113-14, *167*, 176,
    *178, 193*, 210-11, *210, 216*
Martin-Marietta, 114
materials science, 79
"Mathematica: A World of Numbers . . .
    And Beyond" (museum exhibit),
    224
mathematical computation, 123-24,
    137, *146*, 149
    electromechanical, 126, 146, 150
    electronic, 145-46, *150*, 153, 161
mathematical symbols, as design
    elements, 146-47, 148
mathematics, theoretical, 157

Mauchly, John, 39-40, *40*, 41, 126, 131, 133, *135*, 154, 233
McCulloch, Warren, 207-8
McMahon, John, *174*
measurement, electronics and, 47-48, *47*
medical electronics, 46, *52*
Megaw, Helen, 77
Meissner organ-piano, 57
Melpar Electronics, *85*, 96, *96*
memistor, 215
Mercury, Project, 168-69, 171, 172, 181, *181*
Mercury 7, 216
*Metropolis* (film), 198-99
microchips, 119-20, 123
microcircuits, 24
microelectronics, 119-21
microminiaturization, 87
microphone, electronic, 201
microphotography, 20, 76-77
Million Random Digits project, 168
Mills, Neil, 167, 168
mind, vacuum tube as visual metaphor for, 38-39
Mindell, David, 178
Mindell, Joseph, 237
miniaturization, 10, 63-64, 85, 87, 119, 141, 161
    and invisibility of modern electronics, 11, 24
    micro-, 87
    of space electronics, 177-78, 181
Minsky, Marvin, 215, 236
*Missiles and Rockets*, *85*, 94, 96, 146, *167*, *177*, *181*, *186*, *193*, *202*, *204*
MIT, 20, 79, 157, 162-63, 193, 213
    Media Laboratory of, 162
    *see also* Lincoln Laboratory
MIT Press, 162
modernism, 16, *21*, 23-25, *83*, 89, 90, 113, 119, 136-37, 201-3, 215, 216
modernity, electronics and, 23, 86
Moll, Thérèse, 162

Mondrian, Piet, 89
Monrobot, 210
Monroe Calculating Machine Company, 210
Moore, Gordon, 117, 120
Moore's Law, 120
Mora, Raul Mina, *47*, *104*, 105
Morse, Samuel, 123, 137
Morse code, 34, 123, 137, 139, 177
Morton, J. A., 85
movies, 56
    robots in, 198-99
Mumford, Lewis, 106-7, 108
Murarama, Tomoyoshi, 119
Murch, Walter, 227
Murch, Walter Tandy, 75-76, *75*, *124*, 125, 227
Museum of Modern Art (New York), 32
music:
    "canned," 199
    electronics and, 56-57, 59
    punched cards and, 136-37

Nancarrow, Conlon, 136
NASA, 115, 126, 151, 172, *174*, 175, 178, 180, *183*, 186, 189, 216
    Artist-in-Residence program of, 185
    Dawn spacecraft of, 188
    Electronic Research Center of, 186, 188
    *Ranger 3* spacecraft of, 174-75, *174*, 186, 190, 238
    *Voyager* space probe of, 191
National Bureau of Standards, 105
National Cash Register Company, 151, *152*
natural world, integration of technology and, 32
Navy, U.S., 178
neoclassicism, 75
NERVA rocket, 189, 190
Neumann, John von, 41, 145, 150, *150*, 204-5

neural computing, 213

neuristor, 215

*New Landscape in Art and Science, The* (Kepes), 20

*New Yorker*, 81

Nobel Prize in physics, 87

NORAD (North American Air Defense), 64-65

North American Aviation, 176

Noyce, Robert, 117

nuclear energy, as power source, 42, 190, *196*

nuclear force, 69

nuclear weapons, 42, 190
  testing of, 190, 191

*Nucleonics*, 42, 189, 217

number poems, 167, *167*, 168

number strings, 167-69

N. W. Ayer, 30, 31

O'Gorman, Ned, 56

ondes Martenot, 56-57, 59

orthoscopes, 51

oscilloscopes, 46-48, *47*, 56, 65, 225

Overhage, Carl F., 237

*Ozma of Oz* (Baum), 198

Paglen, Trevor, 193

panopticon, 65, 77

Papert, Seymour, 215

paper tape, 125, *125*, *130*, 136, 137-39
  as design motif, 139, *139*, 146-47, *149*, *158*, 159, *206*

Parrish, Maxfield, *12*, 13

pattern recognition, 157-58, 213, 215

PCM (pulse code modulation), 131, *133*, 138, 183-85, *186*, 208

Pennsylvania, University of, 40-41

Perceptron, 213, 215

"persistence of vision," 55

Petrel missile, *207*

Petruccelli, Antonio, 18-19

Philco, 110

*PhotoFact Reporter*, 101, *101*

photograms, 128

photography, 24
  graphic art and, 141, 161, 219
  micro-, 20, 76-77
  printed circuits and, 107

photolithography, 107, 117

π (Pi):
  calculation of, 145, 146, 167
  as design element, *148*

Pierce, John R., 183

Pietenpol, W. J., 85

piezoelectricity, 71, 72, 75-76, 79

planetary model, electronics and, 220-21

player pianos, 136-37

*Popular Science*, 105

posthumanity, 209, 215

postmodernism, 219

postwar era, 24

*Powers of Ten* (film), 224

Princeton University, 41

printed circuits, 105-11, *106*, *107*, 113-16, 135
  in computers, 109-10, 141
  flexible, 115-16
  master drawings for, 107
  *see also* integrated circuits

*Prisoner, The* (TV show), 232

P. R. Mallory & Co., *139*

*Problems of Cybernetics*, 205

*Proceedings of the Institute of Electrical and Electronic Engineers* (Proc IEEE), 26, *27*

*Proceedings of the Institute of Radio Engineers* (Proc IRE), 26, *26*, 27, 29, *43*, 48, *52*, *63*, *68*, *69*, *72*, *79*, *80*, 81, *82*, *120*, *128*, 150, *154*, *157*, 158, *160*, *165*, *178*, 180, 208-9, *210*, *213*, 215, *215*, 216

programming languages, 147, *150*, 152-55

Pulsa, 20, *118*, 119

pulse code modulation (PCM), 131, *133*, 138, 183-85, *186*, 208

punched cards, *39*, 121, 124, 133-37, 141
    as design motif, 125, *125*, *134*, 148, 159
    music and, 136-37
    textile manufacturing and, 133-34, 135
punched cards and, music, 136-37

quantum physics, 68
quartz crystals, 70-71
    in clocks, 73-74, 86
quipu, 127

radar, *48*, 49-50, *49*, 162, 166-67, 177
radiation, atomic symbol and, 41-42
Radiation, Inc., 184-85, *184*, *185*, *186*
Radicon tube, 59
radio, 10, 29, 79, 110
    AM, 29-30, 55-56
    circuit boards for, 104-5
    FM, 29-30, 37, 41, 55-56, 223
    growth of, 14-15
    pulse code modulation and, 131
    transistors and, 88
    wartime applications of, 33-34, 73
radio astronomy, 50, *160*, 172
*Radio-Electronics*, 22, *107*, 110, *110*
"Radio Uses of Piezo-Electric Crystals," 67
Ralph M. Parsons Company, *46*
Ramo, Simon, 171
Rand, Paul, *126*, 128-29, *131*
Rand Corporation, 168, 213
*Ranger 3* spacecraft, 174-75, *174*, 186, 190, 238
Rangertone organ, 57
Raytheon, *68*, *168*
RCA, *15*, *17*, 18, *23*, 25, 27, 29, *49*, 51, 52, *52*, 57, 58, 59, *63*, 64, 88, *98*, 105, *106*, 110, 150, 173, 185-86, *186*, 200, *200*, *218*
Red Scare, 114
Reeves Sound Laboratories, 73

Remington Rand, *40*, 41, *135*, 234
Remington Rand-UNIVAC, 183, 184, 234
Republic Aviation, 188
resistors, 88, 103
robotics, robots, 125-26, 201, 216
    cyborgs as mirror images of, 197-98
    graphic art and, 201-3, 205-6
    in movies, 198-99
    science fiction and, 197, 198, 203, 209
    space program and, 171-72, 173, 174-75, *174*, 176, *182*, 186, 188, 190, 191, 197, 210, 218
    in warfare, 218-19
Romanticism, 95
Rosenblatt, Frank, 215
Rover, Project, 189
Royal McBee, *150*, 154
Rubylith masking film, 107-8
Rutherford, Ernest, 31, 35-36, 171

Sagan, Carl, 191
SAGE (Semi-Automatic Ground Environment) System, 162, 166
*St. Louis Post-Dispatch*, 133
St. Regis Circuits, *105*
Sandia Corporation, *70*, 77, 79, *79*
Santa Fe, N.Mex., 190
Sargrove, John, 105, 110, 119
satellites, 115, 126, 176, 191-92
    Corona series, 177
    *Echo* series, 183-84, 185
    *EchoStar, XVI*, 193
    espionage, 114, 177, 185-86, 191
    *Explorer*, 175, 240
    as extension of human sensorium, 191
    IGY and, 173, 175, 176, 240
    as space junk, 191
    *Sputnik*, 115, 172, *180*
    telecommunication, 183-85, 191, *198*
    *Telstar*, 184-85, *185*

TIROS, 186
transistors in, *175*, 177
vacuum tubes and, 175
*Vanguard*, *175*, 177, 240
weather, *182*
Saturn rocket, *172*, 186
scale, human sense of:
lightbulb and, 13
telephone and, 13
schematic symbols, 92-96, *92*, *93*,
98, *100*, 101
Schenectady, N.Y., 29, 33
Scherschel, Frank, 19
science, dialogue between art and,
20, 21
science fiction, 189, 195
bionics and, 197, 198, 217
graphic art of, 24, 174, 203
robotics and, 197, 198, 203, 209
*Scientific American*, 15, *17*, 22-23, 26,
*39*, *46*, *57*, *64*, *68*, *70*, 75, *75*, 76,
*79*, *136*, *150*, *152*, *172*, *178*, *184*,
*185*, *206*, *218*
Scott, Al, 36, *36*
Selfridge, Oliver G., 152
semiconductors, 82-83, *82*, *83*,
86-87, 88, 110, 117
*see also* transistors
sense perception, 205
*see also* human sensorium
Shanks, Daniel, 145
Shannon, Claude, 131, 133, 138, 184,
203-5, 208
Shelley, Mary, 67, 198
Shiprock, N.Mex., 117, 190
Shockley, William, 86, 87, 88, 89, *91*,
117, 228
Shockley Semiconductor, 89, *91*
Shuttle program, 176
signal amplification, 86, 88
silicon, 86, 89, *90*, 117, 223
Silicon Valley, 89, 117
silk-screening, 107
sine-wave generator, voltage-
controlled, *118*

*Six Million Dollar Man, The* (TV show),
195
slide rule, as analog computer, 127,
231
Sloterdijk, Peter, 223-24
Smith, Cyril Stanley, 77
Smith, Ken, *178*, *188*
snowflakes, 76-77
solar system, as metaphor for
electronics, *11*, *12*, 13, 23, 31, 36,
37, 220
solid state physics, 99-101, 163
sonar, 37, 71-72
Soviet Union:
computer science in, 205
hammer and sickle icon of, 94-95
*see also* Cold War; space race
*Space/Aeronautics*, 108, *109*
space electronics, 24, 114-15, 120,
171-93, 217-18
computers in, 177-78, 180, *180*
electric propulsion and, 188-89
human spaceflight programs, 168-
69, 171, 172-73, 176, 178, 181-82,
185, 191, 195, 197, 210, 216, 217
military programs, 171-72, 173
miniaturization of, 177-78, 181
radio astronomy and, 50, *160*, 172
science-driven programs, 171, 172,
173, 174-75, *174*, 176, *182*, 186,
188, 190, 191, 197, 210, 218
telecommunications and, 183-85,
191, *198*
vacuum tubes and, 175
*see also* satellites
*Space Images* series (Bisttram),
190-91
space race, 114-15, 151-52, 172-73, 177
space-time continuum, 85-86, 171
speech, synthesized, 57, 147, 155,
158-59, 213
*Sputnik*, 115, 172, *180*
spy satellites, 114, 177, 185-86, 191
Staley, Ken, 79, *80*, 81
Stanford Research Institute, 89

Steele, George, 143
Sterling, Bruce, 130-31
Stokes, Warren, 18-19
Stokowski, Leopold, 59
Strategic Air Command (SAC), 65
Stromberg-Carlson, *60*, 61, 72, *72*
*Structure in Art and Science* (Kepes, ed.), 77
*Studies for Player Piano* (Nancarrow), 137
*Suburban Life*, *14*
Sudaka, Gaston, *204*
surrealism, 15, *52*, 63, 75, *202*, 203
surveillance technology, 51, 63, 64, 65, 114, 151, 162, 166, 205, 219
Swiss graphic tradition, 162-63
switch systems, 15, 125-26, 199

tablet computers, *130*
Teal, Gordon, 89, *90*
technics, 106-7
"Technological Impact of Transistors, The" (Morton and Pietenpol), 85
technology:
    change and, 117-18
    as dehumanizing force, 136
    dialogue between art and, 20-25, 27, 96, 148, 163, 173, 219
    integration of natural world and, 32
telecommunications, 10, 15, 71, 79, 105
    automation of, 123-24
    pulse code modulation and, 131
    space electronics and, 183-85, 191, *198*
telegraph, 123, 137
telemetry, 184-85
telephones, 12-13, 79
    and human experience of distance, 13, *14*
telephone switches, 15, 125, 126
telepresence, 218
television, 10, 29, 30, 45, 46
    brain compared to, *201*

circuit boards and, 110
    CRTs and, 50-52, *52*, 55-56, *55*
Televox, 126, 199, 213
*Telstar* satellite, 184-85, *184*, *185*
Tesla, Nikola, 13, 126
Texas Instruments, *87*, 89, *90*, 116
textile manufacturing, punched cards and, 133-34, 135
theremin, 56-57
Theremin, Leon, 56
Thomas Alva Edison Foundation, *35*
Thomson, J. J., 35, 69-70
*Thoughts on Design* (Rand), 128, *131*
Tik-Tok (char.), 198
time, human experience of:
    Einstein's discoveries and, 85-86
    lightbulb and, 13
time-based media, 56
TIROS satellite, 186
Titan rocket, 114-15, 210
Toffler, Alvin, 143
trade literature, 22, 26, 37, 108
TRADIC, 99
"Transistor Fundamentals and Applications" (RCA booklet), *98*
transistors, 24, 41, 83, 85-101, *87*, 103, 116
    computers and, 99, 110, 141, 145-46
    discovery and development of, 86-89, *86*, 208
    graphic art and, 86, 89-96, *91*, *92*, *93*, 98
    in radios, 88
    in satellites, *175*, 177
    schematic symbol for, *92*, 93, *93*, 95, *100*, 228-29
    silicon-based, 89, *90*
    vacuum tubes vs., 87-88, 98-99, 117
transportation networks, switches for, 15, 125-26
*Tubes and Transistors: A Comparative Study* (GE booklet), 98-99, *99*
Tukey, John, 133

Tung-Sol Electric, Inc., *47, 100, 136*
Turing, Alan, 203
*2001: A Space Odyssey* (film), 158–59
typographic art, 23, 77, 143, 147–49, *147, 148*, 151, 169
    Bauhaus School and, 148
    integration of images with, 148, 162–63
    Swiss tradition in, 162–63

Ulano, 107–8, *108*
Underwood, *140*
Unger, Stephen, *157*, 158
Unicode, 139
United Nations, Outer Space Treaty of, 191
United States, interwar immigration of European artists and scientists to, 15–16, 24, 29, 32, 190
United States Rubber, *204*
UNIVAC (computer), 40, *40*, 41, *135*, 233–34
UNIVAC (corporation), 140, 150, 159, 183, 184, 233–34

vacuum tubes, 13–14, 18, 24, 37–39, 45, 86, 95, 103, *106*, 208
    computers and, 39–41, 110, 126
    as obsolescent technology, 43, 85, 100
    radio and, 14–15, 29–30, 37
    in satellites, 175
    schematic symbol for, 95, 98
    transistors vs., 87–88, 98–99, 117
    as visual icon for electronic age, 36
    as visual metaphor for human mind, 38–39, 90, 91
    World War II and, 34, 36
*Vanguard* (IGY satellite), *175*, 177, 240
Vardo, Peter, 18–19
Varian Associates, *57*
Vela, Project, 190–91
Veterans' Administration, 57
video artists, 56

Viennese Secession, 95
*Viking* landers, 197
Vitro Laboratories, *211*
voltage-controlled sine-wave generator, *118*
*Voyager* disk, 191
*Voyager* space probe, 191, 197
V-2 rocket, 37

war production industries, 151–52
*Washington Star*, 172
Watson Scientific Laboratory, 145
weaving, computers and, 121, 133–34, 135
Western Electric, 76
Westinghouse, 19, 125–26, *126*, 128, 199, 201, 213
*What Technology Wants* (Kelly), 220
Whitehill, Joseph, 103
Wiener, Norbert, 204–5, 208, 209, 218
women:
    as computer programmers, 25, 155, 235–36
    in electronics industry, 25, *82*
*World Geo-Graphic Atlas* (Bayer), 32
World War I, 71, 126
    radio and, 33–34
World War II, 15, 24–25
    aviation communications in, 33–34
    digital computing in, 126
    expansion of electronic technologies in, 41, 205
    radar and, 49
    radio and, 34, 73
    vacuum tube technology and, 34
Wrench, John, 145

X-1 aircraft, 176
X-15 aircraft, 176
x-rays, 68, 70, 200

Yale University, 20

Zuse, Konrad, 39
Zworykin, Vladimir, 50–51